MW01120544

The Man Who Invented Gender

Sexuality Studies Series

This series focuses on original, provocative, scholarly research examining from a range of perspectives the complexity of human sexual practice, identity, community, and desire. Books in the series explore how sexuality interacts with other aspects of society, such as law, education, feminism, racial diversity, the family, policing, sport, government, religion, mass media, medicine, and employment. The series provides a broad public venue for nurturing debate, cultivating talent, and expanding knowledge of human sexual expression, past and present.

Other volumes in the series are:

Masculinities without Men? Female Masculinity in Twentieth-Century Fictions, by Jean Bobby Noble
Every Inch a Woman: Phallic Possession, Femininity, and the Text, by Carellin Brooks
Queer Youth in the Province of the "Severely Normal," by Gloria Filax
The Manly Modern: Masculinity in Postwar Canada, by Christopher Dummitt
Sexing the Teacher: School Sex Scandals and Queer Pedagogies, by Sheila L. Cavanagh
Undercurrents: Queer Culture and Postcolonial Hong Kong, by Helen Hok-Sze Leung
Sapphistries: A Global History of Love between Women, by Leila J. Rupp
The Canadian War on Queers: National Security as Sexual Regulation, by Gary Kinsman and Patrizia Gentile
Awfully Devoted Women: Lesbian Lives in Canada, 1900-65, by Cameron Duder
Judging Homosexuals: A History of Gay Persecution in Quebec and France, by Patrice Corriveau
Sex Work: Rethinking the Job, Respecting the Workers, by Colette Parent, Chris Bruckert, Patrice Corriveau, Maria Nengeh Mensah, and Louise Toupin
Selling Sex: Experience, Advocacy, and Research on Sex Work in Canada, edited by Emily van der Meulen, Elya M. Durisin, and Victoria Love

The Man Who Invented Gender

Engaging the Ideas of John Money

Terry Goldie

UBCPress · Vancouver · Toronto

22 21 20 19 18 17 16 15 14   5 4 3 2 1

Printed in Canada on FSC-certified ancient-forest-free paper (100% post-consumer recycled) that is processed chlorine- and acid-free.

LIBRARY AND ARCHIVES CANADA CATALOGUING IN PUBLICATION

Goldie, Terry, author
The man who invented gender: engaging the ideas of John Money / Terry Goldie.

(Sexuality studies series)
Includes bibliographical references and index.
Issued in print and electronic formats.
ISBN 978-0-7748-2792-8 (bound). – ISBN 978-0-7748-2794-2 (pdf). –
ISBN 978-0-7748-2795-9 (epub)

1. Money, John, 1921-2006. 2. Sexology. 3. Sex (Biology). 4. Sex (Psychology).
I. Title. II. Series: Sexuality studies series

HQ60.G65 2014   306.7   C2014-902402-9
C2014-902403-7

Canadä

UBC Press gratefully acknowledges the financial support for our publishing program of the Government of Canada (through the Canada Book Fund), the Canada Council for the Arts, and the British Columbia Arts Council.

This book has been published with the help of a grant from the Canadian Federation for the Humanities and Social Sciences, through the Awards to Scholarly Publications Program, using funds provided by the Social Sciences and Humanities Research Council of Canada.

UBC Press
The University of British Columbia
2029 West Mall
Vancouver, BC V6T 1Z2
www.ubcpress.ca

This book is dedicated to the memory of David Reimer.

I doubt he would have liked that:
he considered John Money to be the source of most of his woes,
and this book in many ways attempts to restore some balance to the general opinion of Money. I am all too aware, however, that in doing the latter
I have given short shrift to the pain of Reimer's life.

Whatever one thinks of Money,
his work was primarily devoted to helping people such as Reimer,
people who, for one reason or another, were unable to fit
society's expectations of what is "normal" in sex, gender, and sexuality.

Contents

Acknowledgments

Thanks to the Social Sciences and Humanities Research Council of Canada for funding to support research for this book.

My first acknowledgment is to Darcy Cullen. It would be impossible to have a more committed and diligent editor. When there were difficult moments, she was always the perfect support. Many former colleagues and friends of John Money were very helpful. My greatest debt is to Greg Lehne, host and resource without parallel. He invested a lot of energy in this book, and I only hope it is at least somewhat worth that investment. I also offer my sincere thanks to Milton Diamond, Jim Geddes, Pamela Gordon, Richard Green, Dan Herzog, Sally Hopkins, Charles Ihlenfeld, Tom Mazur, Claude Migeon, and June Reinisch, who personally offered many helpful comments and recollections.

I received significant research support from the Hocken Library, University of Otago, New Zealand; Kinsey Institute Library, University of Indiana; and the Alexander Turnbull Library in the National Library of New Zealand.

I have had many research assistants along the way and extend great thanks to all of them. By far the most important was Stephanie Hart, who was with me every step of the way for the research of the book and did excellent work for me on her own both in New Zealand and in the United States. This book would not exist without her. I owe a smaller but similar debt to Angela Facundo, Hannele Kivinen, Charlotte Pederson, and Jonathan Vandor.

Thanks to the University of Regina Press, publisher of *Outspoken: Perspectives on Queer Identities*, edited by Wes D. Pearce and Jean Hillabold, for allowing me to use material from my article "Gay, Straight, and In-Between: John Money on Homosexuality," which appears here in a very different form as Chapter 6.

Finally, a very special thanks to Jonathan Crago and Deanne Williams, who read the manuscript in very different ways at very different times but were essential to the success of the book.

The Man Who Invented Gender

Introduction

Sexual Behaviour in the Human

Anyone interested enough to read this book will probably recognize the subtitle of this introduction: it refers to Kinsey's two volumes, *Sexual Behavior in the Human Male* (1948) and *Behavior in the Human Female* (1953). Did the earth move? Yes, it did. But you will note also that I left off the last word of each volume's title. The earth had to wait for John Money to invent gender.

In 1993, feminist critic Camille Paglia said, "Dr. John Money was one of my principal intellectual influences when I was writing *Sexual Personae*. He is the leading sexologist in the world today." Although Paglia is certainly given to overstatement, she is not overstating the case here. In the 1970s and '80s, Money's presence was ubiquitous, from interviews in many mainstream magazines to numerous appearances on television and constant quotations in newspapers. His pleas for sexual liberation made him hot copy for journalists and a frequent target for the many conservatives who feared the idea of sexual freedom. His theories at the time were at the cutting edge of changing views on intersex, transsexuals, homosexuality, and all aspects of sex and gender.

Moving forward, both the importance of Money and today's popular negative assessment of his work are suggested by Jeffrey Eugenides's Pulitzer-prize-winning novel, *Middlesex* (2003). The protagonist, Cal, brought up as a girl, seems to be becoming male at puberty and so is brought to the clinic of Dr. Peter Luce, the world's most famous specialist on sex disorders. Luce, a thinly disguised version of John Money, is obsessed with his own theories of gender malleability and is convinced that Cal should remain a girl. In a rather typical narrative flourish, Cal establishes his individuation by refusing the diagnosis and starting his new life as a man. In the end, the avant-garde theorist acclaimed by the world is refuted by the individual who understands his own biology.

Born in New Zealand, soon after the Second World War Money moved to the United States to do a PhD in psychology, writing a dissertation on intersex at Harvard. He accepted a position at Johns Hopkins University Hospital as a psychologist in the endocrinology clinic, where he remained until his death in 2006. This superficial, yet accurate, summary of his life is not the stuff of an interesting biography. His own bisexuality and penchant for constant travel were also not marks of an unusual life. Instead, for John Money, all aspects of his existence were funnelled into his professional life. His last published book was *A First Person History of Pediatric Psychoendocrinology* (2002a), and the implication of the title is apt: he was first and foremost a part of the area of expertise that he developed. The rest of his existence was secondary.

In an important way, this book begins with another: John Colapinto's *As Nature Made Him: The Boy Who Was Raised as a Girl.* Shortly after it was published in 2000, I heard a colleague give a talk on transsexuality in which John Money was presented as a source of evil. This rather surprised me. I had read a bit of Money's work, and, while far from an expert on him, I had thought of him as a central positive force in the move to accept the medical necessity of performing the surgery demanded by transsexuals. I asked my colleague why he felt this way, and he said, "Have you read *As Nature Made Him*?" I had not because, at the time, my immediate research was going in other directions. Then, when preparing my book *Queersexlife: Autobiographical Thoughts on Sexuality, Gender and Identity* (2008), I found myself reading much more by Money. When that research was finished, I finally read *As Nature Made Him.* And I was immediately incensed.

As Nature Made Him is an account of one man's encounter with gender reassignment. In 1965, an identical twin had his penis burned off in a circumcision accident and was raised as a girl under the influence and care of John Money. The case, in its anonymous guise, was called "John/Joan," but the man was later revealed to be David Reimer. As an adult, Reimer rejected the gender reassignment and lived the rest of his life as a male. From the time that Money started writing about it in the late 1960s, the case became central to changes in beliefs about the relationship between the social construction of gender and biological sex. It also became central to his fame. Money's own publications consistently used the case as proof of a variety of ideas. But the assumption that he advocated "changing boys into girls" became central to many of the attacks on Money's theories.

According to Colapinto's book, the gender reassignment of Reimer never had any medical validity. In Colapinto's view, the procedure was instituted by Money to justify his ideas but had the consequence of destroying David Reimer. When it was revealed that "Joan" was once more "John" and that Reimer denied that he had ever felt any fit with his female identity, all the conservative views of biology and gender – and all the forces opposed to Money – came together in a "perfect storm." The power of that storm remains unabated – always interrupting, commenting on, and, perhaps, eventually defeating a fuller portrait of Money's work that the present book hopes to convey. Widely referred to in both popular and scholarly publications, *As Nature Made Him* is often used to justify the defeat of John Money's ideas.

Although Colapinto seems to be attacking Money's work as a clinician, he was more interested in attacking his theories, which is actually appropriate, as Money's influence was achieved through his writing rather than through his clinical work. Even in the Reimer case, his treatment of Reimer was of minor importance in comparison to the influence of his theories about the malleability of gender – theories that Reimer's parents actually embraced in response to Money's appearance on television and long before Money had any contact with Reimer. Money was very important to certain patients, but the actual number of patients he encountered was limited, considering his many years in the hospital. While most patients remember him fondly, as noted grudgingly by Colapinto in *As Nature Made Him,* his colleagues tended to find him difficult. Thus, he had less support from surgeons and other medical personnel that would have enabled him to develop broader contacts with patients. Some of his students and postdoctoral fellows, most notably internationally known researcher in gender and sexual development Anke Ehrhardt (professor of medical psychology at Columbia University) and psychologist Greg Lehne (assistant professor of psychiatry at Johns Hopkins), remained close to him throughout his career, but most of them did not stay close to him once their fellowships were over. At the beginning of his career, he did develop close relationships with mentors such as Lawson Wilkins, a pioneer of pediatric endocrinology, and colleagues such as Johns Hopkins endocrinologist Claude Migeon. However, for the most part, Money, whose professional qualification was "clinical psychologist," could be categorized as a somewhat isolated thinker. Throughout his career, what Money did on a day-to-day basis seems to have had limited effect, regardless of what he or others said about his actions. On the other hand, his ideas, and the way they were expressed in many books and articles he wrote, had great impact. That is what I concentrate on in this book.

Money certainly considered himself to be the source of the concept of "gender." As my discussion shows, however, Money might have just popularized a usage that already existed, at least in some small way. Still, this book suggests many much wider issues. Feminism would soon embrace *gender* as a term for the social construction of masculinity and femininity, as opposed to the biological term *sex*. Much of feminist thought, particularly in the 1970s, treated gender as completely independent of biology, as though the history of an almost universal force called the patriarchy had compelled the division of humans into male and female, with all aspects of our cultures the products of that compulsory separation. This was not Money's view of gender, but it continues to define much of the usage. Thus, it must be somewhat surprising that this term could be attributed to a man. On the one hand, this justifies Money's not-small claims to being a feminist. Some have gone so far as to depict Money as opposed to masculinity, which was far from the case. On the other hand, Money's "invention" reflects the power and even arrogance of his approach to science. He tended to assume his own originality or at least innovation in every endeavour. He was certainly a dominant male, and in his interactions with his colleagues he had all the assurance of the stereotypical medical doctor of the time, regardless of the fact that he was a psychologist. While Money's writing often asserted that a medical "truth" can have a brief life, and his manner with his patients was benevolently paternal rather than dogmatically patriarchal, his colleagues and subordinates have many tales that show he did not suffer disagreement and could be verbally abusive in opposition. *Imperious* might be an appropriate adjective.

Expressing Gender: Language and Invention

This book is not a biography, and, in the second chapter, I address some of the difficulties facing anyone attempting to write a biography of the man John Money. Suffice it to say here that a biography about the inventor would likely be a very slim one. In contrast, his "inventions" have a range that animates a fascinating narrative of all aspects of sexuality and gender that dominated discussion in the last half of the twentieth century. From intersex through transsexuality, homosexuality, and sexual liberation, Money's ideas were central. The term *invented* might seem to reach too far, especially given the word *gender* already existed. Yet the definition of *invention* in the *Oxford English Dictionary* (*OED Online*) suggests some of the applicable nuances. To invent is "to come upon, find; to find out, discover." Money's new definition of the word *gender* was but one early example of his lifelong attempts at discovery. He might have "come upon" rather than created the concept of gender, but he certainly discovered or uncovered its possibility and easily convinced many people that it was his word. A second definition of *invent* is "to find out or produce by mental activity." Money's great source of power was his incessant brilliance, but the process of his working life shows that his mental activity could produce too much and led to theories that caused him and others great pain.

For Colapinto, Money's analysis of the Reimer case fit another definition of invention: "To devise something false or fictitious; to fabricate, feign, 'make up.'" And Colapinto would certainly consider that Money was fulfilling another meaning, now apparently obsolete, "to plan, plot." But for Money, the hope of "inventing gender" would have followed our usual contemporary definition of the word: "To find out in the way of original contrivance; to create, produce, or construct by original thought or ingenuity; to devise first, originate (a new method of action, kind of instrument, etc.)." All of Money's theoretical developments, all of his neologisms, show the energy of that aspiration to "original thought or ingenuity; to devise first, originate." All his pride in the ubiquity of the word *gender* showed that glee of origination. And even if he was not truly the founder of the word, he certainly fulfilled another obsolete definition of *to invent:* "To originate, introduce, or bring into use formally or by authority; to found, establish, institute, appoint." Whether he originated *gender,* he certainly used his authority to institute and appoint it.

Money's exploration of intersex in the early 1950s was not the first psychological overview of the topic, but it helped to establish a field that really did not exist before Money arrived to work at Johns Hopkins. Before this time, urologists and endocrinologists worked on what were called "pseudohermaphrodites" and "true hermaphrodites," and there was seldom a general consideration of the psychological and social issues associated with them as persons. Money was also instrumental in establishing the clinic for transsexuals at Johns Hopkins, at a time when there was no other location in the

United States that declared itself a centre for gender reassignment surgery. Beyond his status as faculty at Johns Hopkins, his comments on sexual liberation, paraphilias, homosexuality, and sexual and love relationships in general earned him a reputation as what Paglia (1993) describes as the default source for any questions about sex and gender. He became the expert on sexuality – whether in the popular press in such publications as *Time* or *Playboy* or in the courts. In 1977, the *Johns Hopkins News-Letter* proclaimed that when "Doctor Money Speaks, the Whole World Listens."

Whether one regards Money positively or negatively, his theories of bodies – primarily intersex and transsexual – are his legacy, and so too are his theories of the gender expressions and sexuality of all human bodies. Words and language were a central component of Money's understanding and of his emphatic, at times dogmatic, expressions of that understanding. Even Money's PhD on the psychosocial adaptations of intersex individuals, the expertise on which his entire career rested, was only slightly about his own work with patients, and that part consisted primarily of a record of interviews. The rest was his analysis of the texts of various clinical records. And as has often been said, our ideas must be expressed in language. Thus, Money's obsession with his achievement in establishing that word, *gender,* was about the importance of language in our understanding of bodies and sexuality.

Money here illustrates yet another "obsolete" definition of *invention:* "To compose as a work of imagination or literary art; to treat in the way of literary or artistic composition." While his purpose was always to support research and clinical treatment rather than "art," he was well aware of the importance of creativity in literary presentation. He recognized that both theory and polemic were dependent on rhetoric. Or, perhaps, *inventio*. Plato, in the *Gorgias,* suggests that the various aspects of classical rhetoric are mere mechanics and often deceitful. *Inventio* – the discovery of arguments – might be one of the worst culprits in its love of innovative language. But the word *discovery* suggests the importance of appropriate choice in creating a convincing presentation. There are various points in this book where I question Money's *inventio,* especially when he created arguments that go beyond what the evidence can justify. This is particularly an issue when he draws on his limited knowledge of Australian Aboriginal culture to support his theories of sexual education. But this is also the case for Colapinto's *inventio* in *As Nature Made Him*. The many attacks on Money were based on what happened to David Reimer – a person, a human being – but it is the *inventio* in *As Nature Made Him* – the form of the argument – that created the impression of John Money as some kind of mad scientist.

Sexology and Science: Theories and Expertise

What does it mean to label Money a "sexologist?" He usually referred to his own work as "sexology." Although he certainly liked the scientific flamboyance of the term *pediatric psychoendocrinologist,* during most of his career he seemed less happy with the broad category of "psychologist" than with the broad – and often undefined – category of "sexologist."

The first usage of *sexology,* according to the *OED,* occurred in 1867 in *Sexology as the Philosophy of Life: Implying Social Organization and Government,* by Elizabeth Osgood Goodrich Willard. In this book, Willard expresses the Christian fear that sexual activity destroys the self, the family, and everything else – exactly the attitude Money so vociferously attacked in *The Destroying Angel* (1985b). Willard's comments tend to the global and censorious: "I am of the opinion that a man who uses tobacco is not fit to be a husband or a father" (1867, 390). William H. Walling's *Sexology* (1902) is not that different, although Walling, a physician, wrote the book ostensibly to offer medical help. An early example of what has been called the "marriage manual," it begins with pages of praise and endorsements from clergymen. Although less freewheeling than Willard's book, the attitudes are similar: "In describing the evils of sexual excesses and unnatural practices, we point with the finger of authority which they dare not despise, at the deplorable consequences involved, consequences which none may escape" (1902, i).

In other words, the term *sexology* began with a pejorative connotation that today seems somewhat odd. Yet, by 1913, the term *sexologist* was used more or less as it is today. The *OED* provides the following citation: "1913 W.J. Robinson *Pract. Treat. Causes, Symptoms & Treatm. Sexual Impotence* xiii. 93. Some sexologists claim that in the strict sense of the term there is no such thing as a physiologic pollution." It is interesting to note, however, that the early sexologists – from Krafft-Ebing to Hirschfeld through to Havelock Ellis and Freud – were physicians. From the beginning, the field had a clear medical emphasis, although the sexologist almost invariably researched psychology or psychiatry. By the time Money entered the field in the 1950s, however, medical certification was less and less the qualification of the sexologist. For example, very few urologists who work on sexuality today would call themselves sexologists. Instead, the term came to be embraced primarily by psychologists or others whose qualification is in some sense psychology, such as educators and social workers. This has been important both to the field and to Money. Although the absence of physicians made sexology more open, it also made it seem less prestigious and less precisely scientific. At meetings of the Society for the Scientific Study of Sexuality, for example, there is often an undercurrent of yearning for prestige and scientific respectability. And a general absence of medical doctors.

Money's own approach to his work was both personal and distant. He had enormous personal investment in all aspects of his professional life, and yet, throughout his writing, he tried to maintain an air of scientific detachment. This is most evident in the contrast between his own representation of the Reimer case and that provided by Colapinto. To Colapinto, Money was a medical demon, manipulating all around him to his own benefit. However, Money's many references to the case offer little suggestion that he was involved as anything other than a keen observer. Returning to the title of his book, *A First Person History of Pediatric Psychoendocrinology,* mentioned earlier, rather than employing *first person* as a narrative stance – a reflective subjectivity watching – Money presented himself as a person who was there first and thus best able to provide an account of this field with the very medical, multisyllabic name. For Money the thought process was as it is for the stereotypical scientist: the thinking produces the knowledge, but it is the knowledge, rather than the thinking, that counts.

Just as Money the psychologist-sexologist represents a change from the early sexologists, so too does he represent a change, in various ways, from his more immediate precursors. In the field of intersex, one might look to a urologist such as Hugh Hampton Young or an endocrinologist such as Lawson Wilkins, but neither pursued the general category of intersex in the manner that Money did. Albert Ellis, another psychologist who wrote paeans to sexual freedom, wrote an article that provided a general and generous look at intersex in 1945, but he never built on it. While Harry Benjamin was more of an activist in transsexuality than Money, he lacked the entrepreneurial skills that allowed Money to set up the Johns Hopkins clinic. Kinsey's group certainly did much more methodical and extensive research on human sexuality than Money. Although Money laboured ardently to maintain longitudinal studies on the intersex children who came to Johns Hopkins, most of his theories on sexuality and gender were based on very limited clinical data. His ideas tended to be rather inductive and were then checked out via the literature. The concepts were not predominantly from work with his own patients. On the other hand, he was ready to make general theories on all aspects of sexuality and gender. Some of his most brilliant and perceptive ideas were produced on sparse experience, which makes him, in a sense, an anti-Kinsey, given that Kinsey tended to see his own primary contribution more in acquiring and arranging data than in theorizing from those data.

Money is similarly different from William H. Masters and Virginia E. Johnson, or, as they are forever known, "Masters and Johnson," who were first and foremost clinicians and therapists with extensive laboratory studies. Their published work tended to emphasize aspects of behaviour and physiology that needed either change or reinforcement. Money's work in similar areas, most obviously in his theorization of the lovemap, used little or no laboratory information and had limited interest in physiology. Instead, Money was primarily concerned with the ideas that he gleaned from studies that emphasized psychology. One might go so far as to suggest that he presented a philosophical response to issues of gender and sex. This approach might seem similar to the later work of Albert Ellis, but, although Ellis certainly had claims to the category of sexologist, his publications might be better seen as theorizations in the aid of psychotherapy, with a concentration on sex. Ellis was never as general or even as flamboyant as Money. While Ellis was controversial and far from retiring in presenting his views on subjects such as his rational emotive behaviour therapy, Money's canvas was larger. He ardently pursued a role as a sex liberationist and was forever in search of a public venue for his ideas. Another academic similarly obsessed with science might have found it off-putting to attempt to place those ideas in a form that would appeal to the popular press, but Money leapt in where other angels of the medical world would quite sensibly have feared to tread. The obvious comparison would be with someone such as Havelock Ellis, who clearly wished to further the liberation of the various sexual possibilities and yet constantly retrenched partly because of fear of censorship but also because of a concern that his ideas not be rejected by the voices for propriety. Money, on the other hand, seemed to believe that, once he established his expertise and bona fides, he should be allowed to say whatever he pleased. And then he need only wait until the world was convinced.

Money was very much a man of his time. His first published words could not have appeared other than immediately after the Second World War, and his last writings, at the end of the twentieth century, were too late, in that they both theoretically and temperamentally harked back to the earlier rise of sexual liberation and a rather freewheeling sexual science. Both of the latter had seemed, in the 1970s, to be an inevitably expanding part of the future, and yet, by the 1990s, a new conservatism had overcome the liberation and was turning the science into a restrained form of analysis that had no room for Money's creative generalizations. The rise of AIDS was not central to any of Money's primary areas of interest, but it had helped to sponsor, or at least coincided with, a repressive morality that had no interest in sexology other than as part of disease prevention. Such a time of restraint was not conducive to Money's theories or to Money's mode of operation. While Money constantly extended and developed his theories in all his areas of endeavour, he seldom changed them significantly. Part of the reason his ideas might seem avant-garde in the 1950s and out of date in the 1990s is that they are much the same in both periods. Having dwelled on his writings for a number of years, I am struck by how consistent he was. In some ways, this is positive in that he developed theories and then worked within them, but it is also negative: there is little suggestion that the research of others or social transformations did much to change his mind. Thus, while the present study always takes note of the date of publication, there is no attempt to follow chronology. Rather, the process in each chapter is to trace the variety of nuances in Money's thought on the subject at hand. The dates are most notable for how little they reveal about his ideas.

This book examines John Money's thoughts on and achievements in his major fields: intersex, transsexuality, homosexuality, and sexual liberation. It does not attempt to trace the fields more than to suggest how his ideas fit within them. While I am always aware of the larger contexts, this book is very much about one man's thoughts rather than about the overall developments in sexology. For overviews of intersex or transsexuality, the reader must look elsewhere, particularly Katrina Karkazis's *Fixing Sex: Intersex, Medical Authority, and Lived Experience* (2008) and Joanne J. Meyerowitz's *How Sex Changed: A History of Transsexuality* (2002). However, Money was a central figure in each of the areas considered here. In some of them, for a time, he was the central figure. In all, he offered theories and made claims that have changed lives and created controversy. He could not have been who he was in another period. The immediate post-Kinsey era of the 1970s and '80s enabled him to operate as a sexologist in ways that Kinsey himself would have thought impossible, such as his many appearances in the popular press. But sexology also would have been a very different field without Money. To become what it has become required his intelligence and imagination as well as his arrogance and often extraordinary hubris.

That hubris might be one reason why most sexologists have given Money slim recognition. Then again, Money seldom referred to the work of others. One exception was his embrace of the word *limerence,* as developed by Dorothy Tennov (1979), which he used often. It was a rarity for Money to take a concept and word created by another researcher and use it so happily. For the most part, when Money mentioned a colleague's work, it was to attack it in letters or reviews. While Money's ideas are important simply because of what they say about sexuality and gender, they cannot be said to be a part of a coalescing of thought that developed through a number of theorists and researchers. Money was forever alone. Those who today extol his achievements are few partly because of that solitude: he did not have many intimate co-researchers and followers who might further his ideas and reputation.

This is ultimately a book about texts. The texts develop from, shape, and comment on something very material – the human body – but they are texts. I often quote Money at length to give full space to his argument, and then I dissect the quotation to show the insights and lacunae as he himself considered the aspects of sex and gender that most compelled his interests and arguably most compelled the interests of his society. References to personal communications with some of Money's patients, colleagues, and friends, as well as my various personal reflections, are used to aid my interpretation of the texts. In my personal reflections, I supplement my knowledge as an academic with what could be called a more general response, something akin to what a literary critic might call the view of an average reader. In some instances, my comments present the specific understanding of a sixty-three-year-old gay man whose life experience has been bisexual. I have found that presenting the views of someone such as Money without reflection makes it difficult to assess those views. My hope here is to consider Money's ideas as carefully and as fully as I can. In contrast with *As Nature Made Him*, which claims to present an objective portrait of a controversial subject and merely uses this to hide an extremely subjective point of view, I have often placed my subjectivity at the fore.

Inventing an Order: The Chapters

The order of the chapters in this book attempts to follow a logical accretion as each chapter provides information that is necessary to understand fully the subsequent chapters. The subtitle of these few opening pages modifies Kinsey's, but the titles of the other chapters are taken directly from other texts. Most are books written by Money himself, but some are by others that in some sense provide a context for Money and his work. Thus, the second chapter, about Money's early years, is titled "Once a Brethren Boy," taken from Noel Virtue's 1996 memoir of the same title about growing up in New Zealand. This chapter shows the formation of what Money would later become. Chapter 3, "Fixing Sex," from Karkazis's title, traces Money's theories of and work with individuals with intersex conditions – his first professional experience and always the core of his views of sex and gender. The fourth chapter, "Lovemaps," uses a term Money created for his theory of how sexuality and gender combine in all humans; the term also appears in the titles of three of his books. This theory developed out of Money's intersex studies, but it represents his attempts to create theories applicable to all human conditions and provides an overview of his methods. As his colleague Greg Lehne has said, understanding lovemaps is a key both to recognizing the sources of Money's ideas and to seeing why they took the shapes they did. The fifth chapter, which is about transsexuality, is called "Man and Woman, Boy and Girl," after Money's most famous book, which he wrote with his then student Anke Ehrhardt. Money's work on transsexuality includes his greatest early professional success – the creation of the Gender Identity Clinic at Johns Hopkins – and one of his earliest sources of professional despair – its closing. Chapter 6, on homosexuality, is called "Gay, Straight, and In-Between" – the title of the book by Money that is most specifically about homosexuality. Money originally thought that homosexuality would be the topic of his dissertation; he arguably made his greatest personal investment in this area, although he seldom acknowledged this. Homosexuality was never more than a tangential interest in his career, but it is the one field in which he had not just professional experience. The chapter on Money's various attempts at neologisms, Chapter 7, is called "The Edge of the Alphabet," the title of a 1962 novel by Janet Frame. This chapter might seem like the orange among the apples, but as was asserted earlier language was always a key to Money's understanding. Chapter 8, on sexual liberation, takes its title from John Heidenry's history of the 1960s and '70s, *What Wild Ecstasy* (1997). As Money became a spokesman for sexual freedom, he used all of his professional expertise in all these areas to justify a highly developed public persona as he went from sexologist to proselytizer of sex. Chapter 9, "As Nature Made Him," the title of which is borrowed from Colapinto's book, focuses on the David Reimer case. Arguably, this case has had the most significant influence on Money's reputation, but it is far from the progenitor of Money's many ideas. It is, rather, the result of them, for good and ill. To understand why Colapinto's book is wrong in so many ways requires a broad knowledge of what Money had written in the other areas. The Conclusion, "Venuses Penuses," is again a title from Money; *Venuses Penuses* is a compendium of his articles, published in 1986. Everything about that book, including its extraordinary title, suggests Money's ultimate successes and failures.

Almost thirty years ago, I wrote a book titled *Fear and Temptation* (1989), which prompted a colleague to remark, "Ah, yes, the inevitable autobiographical title." I am not "the man who invented gender," although some of the discussions in my undergraduate classroom make me feel as though I am. When I present my ideas, inevitable and even timeworn, the audience often finds them novel and even offensive. This is a common reaction to issues of sexuality. Some forty years ago, a television program called *The Larry Solway Show* was cancelled because of sexually explicit discussions. Solway wrote a book about it titled *The Day I Invented Sex* (1971). As Solway said, sex had been around a long time, and it seemed unlikely that he had been responsible for any of it, but those around him acted as though it had lain dormant before his program. Perhaps sex and gender are available to be invented again and again and again.

One note about this book and about John Money: often in the text, multiple-authored texts are attributed just to John Money, although the full authorship is given in the Works Cited. This might seem both unfair and inaccurate, but years of working with Money's writing have shown me that anything attributed to "Money and ..." was written by Money, as Money often asserted. Co-authors had to submit themselves to an iron will, and an iron will with an often idiosyncratic writing style, full of neologisms, scientific jargon, and arbitrary analogies to anything that might come to his mind. The exceptions are the co-authored books *Man and Woman, Boy and Girl* and *Sexual Signatures,* which clearly show the presence of Anke Ehrhardt and Patricia Tucker, respectively. Whether in ideas or words, Money was at his best when he listened to the modifications suggested by others. For the most part, he did not. This increased the flamboyance and innovation of both theories and language, but it also led to extremes in both that would forever damage what he attempted to do.

CHAPTER ONE

Once a Brethren Boy

The Early Years

In his memoir, *Once a Brethren Boy* (1996), Noel Virtue describes his youth in mid-twentieth-century New Zealand, much of it spent trying to live within the strict rules of the Plymouth Brethren and yet recognizing his own burgeoning homosexuality. Twenty years before Virtue, John Money had lived a similar experience. A number of accounts of Money, such as John Colapinto's in *As Nature Made Him* (2000), explain Money's ideas as a reflection of the difficulties of his upbringing in New Zealand. It has become somewhat de rigueur to explain a psychologist through his biography, as has often happened to Sigmund Freud. The worst examples are no doubt those that, apparently unwittingly, use Freudian theory to show why Freud's life history warped him into creating mistaken theories. I say "unwittingly": perhaps it is something in their subconscious.

A proper biography of John Money would be a difficult proposition. The late Michael King intended to write the story of his life but soon abandoned the project. I was told he gave up because he was unable to secure any financial support from the usual foundations. The only published results of King's research are an article for the *New Zealand Listener* magazine (1998) and the biographical section of the catalogue of the John Money Collection at the Eastern Southland Gallery, a major provincial art gallery in Gore, New Zealand. Most of the details of Money's early life noted here are from the latter and the brief article "Explorations in Human Behavior," which Money (1991a) wrote for *The History of Clinical Psychology in Autobiography*.

There might have been other reasons why King decided against writing the full biography. It would not be inaccurate to call Money a closeted bisexual. Richard Green has called him a "libertine" (2008, 613). Money made it clear to King that he did not want to dwell on this aspect of his life. This might make him seem like a perfect candidate for an unauthorized biography: plenty of hidden encounters with strange men and women. However, while this side of his life certainly existed, it does not appear to have controlled his life the way most narratives of the closet suggest is the norm. He was well aware of the professional dangers of coming out, which no doubt contributed to his general interest in sexual liberation. Perhaps had he been heterosexual he would have searched for a way to have a professional life and a personal one. Instead, however, he seemed to have only a professional one. His personal life was primarily as an expatriate New Zealander, with close ties to family, many visits to his home in Baltimore by old New Zealand friends, and, of course, his lifelong friendship with arguably New Zealand's greatest writer, Janet Frame.

Money was born on July 8, 1921, in the small community of Morrinsville. Both of his parents were at least technically outsiders; his father came from Australia, and his mother was born in England but grew up in New Zealand. Neither parent was educated beyond primary school, and, as a couple, they led a rather hardscrabble life where money was difficult to come by. Both were fundamentalist Christians. Money's father came from a Salvation Army family, but in New Zealand they both became members of the Plymouth Brethren.

In *Once a Brethren Boy*, Virtue recalls a boy who had killed his brothers to prevent them from being oppressed by the Brethren religion. Virtue describes the father of the boy coming to speak at his family church:

> Collectively, the Brethren placed all the blame on to the boy, not themselves, not even once questioning whether their teachings were wrong and dangerous. They claimed, in their frightening, ignorant and inhuman obsession with righteousness, that the act had been carried out under the control of the Devil. The boy was a backslider and had fallen from grace. I recall no sympathy for him from his father. He has probably spent the rest of his life inside his own earthly hell.

> The events left an impression on me so strong that I cannot ever forget sitting there in that Gospel Hall staring up into that father's unforgiving face, listening to his words. Decades later, my father was to write a letter to my publishers using almost the same words that the boy's father had spoken about his son; that I was also filled with evil and Satan. This was because I was homosexual and had the audacity to write about the Brethren in my first novel, *The Redemption of Elsdon Bird.* (1996, 23)

Money was neither such an out homosexual nor so overtly critical of the Brethren, but he clearly felt that religious repression had a very negative effect on his youth:

> By around age 12, I needed to explain various of the fundamentalist precepts with which I had been reared, but I could not. Nor could anyone else. Belief, I was told, is based on faith, not fact. That did not sit well in a mind that even in infancy had annoyed adults by forever asking What? and Why? It took me another twelve years or more to escape from the philosophy of hellfire and damnation to the philosophy of scientific skepticism. (1991a, 233)

The escape is represented in a paper called "Psychiatry and Religion" that Money wrote while working as a junior lecturer at the University of Otago. He asserted, "The fact is that all those who hold absolutist beliefs are in many ways similar to the paranoiacs with their delusionary beliefs" (n.d., 2).

Money's "Explorations in Human Behavior" begins with a memory of being envious of a circumcised playmate when he was five years old: "I was an early beginner in the cosmetics of sex, for I wanted my own glans penis to be like it" (1991a, 232). Then came school, where he found the battles of his male peers quite overwhelming and went home in the middle of the day:

> My alibi was that I thought the school day was over. I revealed nothing of my monstrous panic and failure as a warrior. I was on my way to becoming a psychological exile who, excluded from the establishment, would survive by getting to investigate it, challenge it, and outwit it. Investigating, challenging, and outwitting – these also apply to my professional research activities in psychoendocrinology, sexology, and psychopathology in general. (1991a, 233)

While he might not have been a warrior, Money responded by becoming what Australians would call a "little battler," the tenacious individual who survives in spite of lack of opportunity. He came from a world in which physical violence was quite normal and translated it into intellectual warfare.

For outsiders, the image of New Zealand as warlike might seem rather surprising. The tourist brochures suggest two idyllic islands where you can swim and climb mountains on the same day, with welcoming attitudes from a multiracial population. The vexed relations with Indigenous peoples in Australia or Canada are met by a country that incorporates Māori in its proud bilingualism and even uses its Māori name, Aotearoa, on government documents. Yet Money recalls "playtime" as "a replay of the Waikato Māori wars" (1991a, 232).

This, of course, is more than eighty years ago and specific to a location with a significant white – or, to use the New Zealand term, *pakeha* – and Māori population, but this undercurrent of violence continues today. Just as one example, according to Ross Kemp in *Gangs* (2008, 45), there are more bikies, what North Americans would call bikers or motorcycle gangs, per capita in New Zealand than anywhere else in the world. Their fortress-like clubhouses in suburban New Zealand are quite amazing. To offer a personal anecdote, I was sitting in an airplane waiting for a domestic flight in New Zealand when a young *pakeha* man came on. He had the word *Mongrel* tattooed on his forehead in gothic letters, which signified his commitment to the gang known as the Mongrel Mob. Such facial markings can be seen throughout New Zealand. Whether in this form or Māori patterns, they are usually believed to be reflections of the *moko* traditionally worn by a Māori warrior.

New Zealand violence is definitely not limited to the Māori or to interracial conflicts. There is no question that the Māori had a warlike history before the invasion from Europe, and such violence continued in the Māori wars in which they fought the English invaders. While racial discrimination is a factor in New Zealand as in most countries, Māori have a social power that is envied by other indigenous peoples. However, a high level of antagonism in New Zealand appears in many contexts with little or no Māori presence. I have been involved in political demonstrations that led to conflicts seldom seen in Canada or Australia. Over the years, I have been in many discussions about this underlying anger: some claim this is a remnant of the Māori history, and some have even claimed it is a result of the economic imperatives of slaughtering a lot of sheep.

This violence is also quite specific to masculinity, however. In *One of the Boys: Changing Views of Masculinity in New Zealand* (1988), a collection edited by Michael King, a multiracial group recalls growing up in a world oppressed by alcohol and male violence, where becoming a man was a dangerous business. Money represented his early life under the title "Serendipity," and he began this section of "Explorations in Human Behavior" with an account of his father's cruelty:

> The abusive interrogation and whipping that my father gave me when I was four had the serendipitous effect of confronting me with a lifetime's rejection of even the type of workman's clothing he wore that day, while demonstrating the brutality of manhood and the moral self-righteousness of authority. My father harassed me until I said yes, that I and not my playmate had broken a glass pane in his new garden frame. In fact it was an accident of two boys' horseplay and I did not know who was the one responsible. My father died without my being able to forget or forgive his unfair cruelty. (1991a, 238)

Money's father died when Money was eight, so there was no opportunity for them to develop a different relationship. However, one might see a resemblance between father and son in their definitive objections to authority. When Money became a conscientious objector in the Second World War, for example, he was following a family tradition. His father had been interned as a conscientious objector during the First World War.

Money was also certainly skeptical of religion and many sorts of "absolutist belief," but he tended to be, if not absolutist, certainly dogmatic in his opinions, although the dogma was invariably of his own devising. Whether this is simply a reflection of the very different kind of dogma of his childhood, it is impossible to say. This rather autodidactic dogmatism no doubt contributed to the various difficulties he had with colleagues. Although Money both founded and led various sexological organizations, he was seldom the person in charge, and he was never in charge at Johns Hopkins Hospital, where he spent his working life. He forever identified as the lone wolf struggling against those who had administrative power.

Money grew up in a predominantly female household. He recalled his spinster aunts as particularly bitter:

> They resented being single, envied their sisters-in-law by denigrating them, and never-endingly railed against the injustices of living in a man's world.

> I suffered from the guilt of being male. I wore the mark of the best of man's vile sexuality. I wondered if the world might really be a better place for women if not only farm animals but human males also were gelded at birth. The antisexualism and the antimasturbation hysteria of Victorianism spread a sinister influence over me and my generation of youth. It was sinister enough to keep me, off and on, professionally and personally engaged in rectifying it over the full course of my career. (1991a, 254)

As Nature Made Him keys on these observations, but the selective quotations invert Money's meaning. One brief example will suffice here. Colapinto ([2000] 2006, 27) states that Money said, "I wore the mark of man's vile sexuality," and he thus sees these words as exemplary of Money's opposition to male sexuality. This is a more than significant erasure. I do not deny that the original – "the best of man's vile sexuality" – is confusing, but it must mean something quite different from what Colapinto claims. One possibility is that this "best" simply refers to his own genitalia. My own reading is that the male guilt was something he quickly and resolutely overcame and evermore devoted himself to "rectifying" wherever it could be found. His rejection of brutal masculinity did not make him reject maleness; rather, it left him exulting in the best of what the aunts viewed as "vile sexuality." One of Money's key moves in opposition to anti-sexualism was to assert the value of male sexuality, including male lust, and to reject the idea that it could not be "the best." He recognized that there could be a vital maleness, with an unthreatened virility, that was not dependent on violence.

Money's childhood after the death of his father was one of consistent poverty. His aunts were present at least partly to help with expenses. If one associates artistic pursuits with genteel leisure rather than scraping for a basic living, it might seem surprising that so much of Money's attention was devoted to the arts, but this is not out of keeping with his family. Although John remembered his father, Frank, as a brutal workman, he was also a fine painter. (Some of Frank's paintings appear on *Moneys Creek,* a website maintained by one of Frank's nephews.) John Money was an amateur artist throughout his life as well as an ardent supporter of painters and painting. He was also an accomplished pianist. As a young man, he aspired to a career as a musician and, according to his comments recounted by various colleagues, was quite devastated when he came to the conclusion he would never be good enough to be a professional.

When Money graduated from high school at sixteen, he was too young to enter university and so spent a year as an "uncertified teacher" before he began at the University of New Zealand in Wellington, where his family was then living. At the same time as he attended university, he went to a teacher's training college, where he received a small stipend. Between free tuition at the university and living at home, it was sufficient to survive. On his graduation from teacher's college, however, his status as a conscientious objector denied him a teaching position. Thus, his only obvious possible future was to continue in the university, and he accepted an appointment as a junior lecturer at the University of Otago in Dunedin.

Money's reminiscences do not make his original purpose at university clear. It seems that he had already put music behind him. His choice of psychology, however, might not have been completely a choice. He makes particular note of the difficulty of acquiring Havelock Ellis's *Psychology of Sex* in New Zealand in 1939, but seeking the Ellis volume could be as much a reflection of his interest in the scientific study of sex as in psychology. He noted that his teacher training classes precluded morning lectures at the university, and thus

> I was obliged to exclude the natural sciences from my degree. Psychology was more or less derided as a science, for it still had not gained independence from the Philosophy Department. Nonetheless, it was the only science I could study as an undergraduate. (1991a, 239)

So perhaps the real choice was science, and psychology was just the only possible science. A letter from 1946 shows that he had a particular aptitude for medical science, as he took a course in physiology while he was employed as a lecturer at Otago, and the professor of physiology seems to have been impressed: "I think he is rather intrigued by the fact that a psychologist is keen on understanding physiology, for he has rather got the idea that psychology is a bundle of mystical ballyhoo." In the same letter, Money suggested he might take a medical degree in the future.

Money's letters record many thoughts of his years at Otago, often about the failures of the philosophical teaching of psychology. He was quite unhappy with a new professor of philosophy, as he noted in a 1946 letter to his family: "He certainly has very little tolerance for psychology as a subject in the university, with the result that there are always little pin prickings and discontents." In the same letter, the discussion of a student prize made Money particularly angry at the professor's attitude:

> This leads him to say perfectly silly things such as that "We must remember that we are in the arts faculty and when marking essays consider literary style." That is perhaps true for philosophy where everything is rather vague and wooly, and depends a lot on personal opinion. But our subject is a science, in which you either know your work or not, and very much easier to give high marks.

Money's commitment to science was visible throughout his undergraduate work. For example, he ended one paper, "The Social Scientist on Freedom" (1944b), with a reflection on a new book that had immediately captured wide attention in the United States – James Burnham's *Managerial Revolution* (1941). Money concluded,

> Perhaps, however, seeing the trend of social change, it will be possible even to plan for freedom, to devise a plan whereby the undoubted values of a rationally planned society can be achieved without a ruthless dictatorship of a managerial class. One envisions a completely independent research body whose job is to plan for rational planning within great society.
>
> Maybe it is a distant dream. More so, the need for dreamers. (1944a, 26)

It is no doubt intriguing that a young man so attracted to a "rationally planned society" should spend the rest of his life researching sexuality, not the most rational of human attributes.

Money combined his psychological studies with his musical interests in his MA thesis, "Creative Endeavour in Musical Composition" (1944a). The study was far from original research. It was primarily reflections on published memoirs and various secondary sources, but the conclusion is perhaps surprisingly representative of Money's tone and attitude throughout his career:

> There is nothing more frightening than ignorance. Then again, so much bunkum has been spread abroad on inspiration that it has the bad reputation of being a substitute for hard work and skill; and so for almost the reverse reason, modern musicians especially have feared to confess it.
>
> Perhaps it is just as well that the job of studying inspiration, or insight as it is called, has been ultimately left to the psychologists. (1944a, iv)

For Money, the "psychologists" were scientists who would eventually remove the "bunkum" of even something so amorphous as inspiration.

In general, these psychologists were American. As in so many other aspects of New Zealand life, the university was very much under the influence of its British heritage. In one interview, Money remembered,

> There was no PhD offered in the University of New Zealand at the time I left in 1947; I believe there had been in around 1880, in chemistry or maybe physics too, and then the senate of the University of New Zealand abolished the granting of a PhD degree because it had high and mighty ideals about keeping up the standards. So when I did my exams for both the Bachelor's and the Master's degree, the questions were all set by unknown examiners in England so that the quality of colonial education would not be allowed to slip. (1986a, 142)

As so often happens in colonial situations, the British form to be emulated was one already rather out of date in its place of origin. In New Zealand, as Money noted, psychology was viewed as far from science. By the mid-1930s, psychology had become part of the Natural Science Tripos at Cambridge. In New Zealand, however, Money found himself moved in progressive directions by a member of the faculty whose influence was not British but American.

Ernest Beaglehole was a devoted New Zealander, and his brother, J.C., is arguably the most famous New Zealand historian. Although Ernest did his PhD in social psychology at the London School of Economics, his academic interests were primarily nurtured by time spent in the United States, first at Yale in the early 1930s and later at the University of Hawaii. His original work in the United States was in child psychology, and his most noteworthy early studies were under the tutelage of Edward Sapir in what many would call anthropology but could also be termed ethno-psychology. Beaglehole took his first position in New Zealand as a lecturer in "mental and moral philosophy" in 1937, but it was not long before he became New Zealand's first professor of psychology. His obituary in the *American Anthropologist* is revealing both in its appreciation of Beaglehole and in its view of New Zealand: "Ernest Beaglehole chose to spend the major part of his life working in New Zealand. He was, nevertheless, a man of international orientation and stature" (Ritchie 1967). His personal and professional connections to the United States impressed the young Money in many ways. The degree of this impression is suggested by the copy of an undergraduate paper, "Delusion, Belief and Fact" (1945) that Money kept in his collection. It has pencilled notes by Beaglehole, some of them quite critical. This might not seem unusual but for the fact that another note on the paper states that Money retraced these marginalia some fifty years later to ensure their continued legibility.

The paper itself suggests Money's attitudes throughout his career. It rejected morality and religion as not provable and instead turned to science:

> The scientific belief is an hypothesis which is capable of a high degree of verification. A highly developed technique of experiment and measurement and a special code of symbols make it possible for the hypothesis to be verified on different occasions and by different people, with always the same conclusion being arrived at. Thus it is highly capable of social communication, irrespective of personal prejudice and idiosyncracy. In so far as these conditions are satisfied it is called a scientific fact. (1945, 7-8)

Money no doubt often veered from the rigours suggested here, but his commitment to the ideal he expresses in this passage never wavered.

One unusual aspect of the John Money Collection at the Hocken Library at the University of Otago is the collection of letters to his family. Many people keep letters, of course, and people who have become somewhat famous tend to treasure those letters to be offered to posterity. In this sense, the Money letters are much the same, beginning with a rather innocuous pencilled note to his "Dad" when he was eight. By his late teen years, however, they changed significantly. The letters were typed, and, in many instances, they bore a note that they were to be passed to all members of the family. A number stated that they were to be returned to the author. By the time he was at Otago, there were no salutations and often no closings. Offhand references in the letters suggested that many in the family offered sarcastic ripostes to this form of correspondence, but Money replied that he was too busy to bother with multiple correspondents or handwriting.

The tone of the individual letters reflects this tenor. This is one:

> "Some Critical Impressions after Working in a Mental Hospital"
>
> My deep interest in the patients as individuals, some of whom have become my friends, and my training in psychology have made me strongly desirous of doing something to improve the conditions which they experience.

> In the first place many undesirable situations arise as a result of the ignorance of the staff concerning fundamental things as to the nature of personality.
>
> Finally I may mention that overseas there is a growing body of psychiatrists who are attacking the problem of psychotic behavior from a psychological point of view, and are having some success. There is considerably greater success with similar treatment for neurotic cases. In the mental hospitals of this country there is no provision for these new therapeutic techniques, nor is their [sic] any provision made to educate young psychiatrists in their use.
>
> J.W. Money MA
> Philosophy Department, University of Otago, Dunedin,
> 9th June 1945

The title and closing would suggest that this piece was meant for publication, perhaps as a letter to the editor, but even the twenty-four-year-old Money would have been unlikely to publish a letter on the "ignorance of the staff" at the local psychiatric hospital. Money's other letters make it more than possible that this was just one more circular to the family. He was always willing to turn these letters into academic lectures.

He was also happy to combine a lecture with jocularity, as in the following letter from 1945, which seems to have been addressed primarily to his sister and to comments she had made about the unscientific nature of psychology and the tendency for psychologists to be obsessed with sex:

> Conversation Not to Have With a Psychologist
>
> You believe, judging from the remarks of a science lecturer whom you once heard at your home-town university, that psychology is not really a science.
>
> You don't need to worry because he knows more about sex than you ever dreamed of, and has achieved such emotional detachment that he can remain neutral, and feel no need to pass judgment.

Sometimes the letter seems to have been intended for circulation but also contains material that attempts a psychotherapeutic intervention for a specific member of the family. While Money's published reference to his aunts presents them as imperiously anti-sexual, this undated letter suggests he was far from intimidated by such attitudes:

> I think that you are inclined to pass judgement on what seems to you immoral. Perhaps there was just a little bit of this in the remarks you made about Glad's latest episode. Why don't you write to me, Pot, you neglectful old aunt, and tell me all about your worries, in the same way as you used to talk to me. I am sorry that there are drives and impulses urging you to integrate love situations with odds and ends of men, when you have had enough experience to know that they always turn out catastrophically. However, perhaps we'll find the meaning behind it all when I am up in Wellington again. I think I know something about such things since so many girls have been behaving in an idiot way with me as the victim. Don't tell me that you can't help it and that it's all their fault; because I just won't believe it.

The most revealing of these therapeutic lectures is from 1946. Addressed to his brother, Donald, it included a wide range of comments on sexuality, beginning with an observation about a friend: "Has he made love to you yet? Seems to me a distinct possibility of his being homo., but admittedly I have very little yet other than superficial indications to go on." He quickly moved to suggest "sex instruction books" for one of his brother's female pupils: "I am quite in earnest about this, for I know only too bloody well from my clinical experience that bad education leads to catastrophe, and it seems to me that with parents who would do such a daft thing as hers have done, there is every chance that she has had a bad show."

One aspect of the letter anticipated Money's later career in endocrinology:

> I have a fairly good grip of endocrinology but I also know only too well that some of the assertions of the endocrinologists are a lot of bullshit. There is a pernicious attitude of mind which assumes that if a physiological basis can be associated with a particular disorder, then the cause must be the physiological basis. This is completely unscientific, and is a relic of the nineteenth century scientific materialistic philosophy. Very recent research into psychosomatic medicine has shown conclusively that many physiological (somatic) symptoms, e.g., gastro-intestinal disorders like peptic ulcer, are very frequently psychologically caused.

Another section of the letter addressed sex and love and was quite similar to Money's public comments in the 1970s, particularly in interviews. It is worth quoting at length partly because it shows Money's consistency but also because so many elements prefigure Money's idea of the lovemap:

> Lust without friendship, those things which give one a community of interests which make a lifetime of sharing in common possible, is just prostitution and as such is nothing more than a milking machine – the satisfaction is about the same. Friendship without lust, when one is trying to make the sharing lifelong, is hopeless and leads to divorce or some other symptom of marital incompatibility. And interestingly enough, if because of one's puritanic bringing up, lust and friendship are incompatible, then the result is likely to be personality breakdown, commonly called nervous breakdown. The most frequent and important single cause of psychological break up is just this. Exactly the kind of thing which happened to your friend at Griffith, whom you mentioned. What is most likely to happen is that the lust factor finally becomes so strong that something happens: that something is love of the electric shock type, the kind of thing that the films have taught us to believe is the only kind of love. But it has no solid foundation, and should the affair turn out badly, then there is complete collapse. It is only what one would expect when all the eggs have been put in one basket, should the basket break.
>
> True love is something which is none the less physically exciting. But there is not the appalling dependence of one person on the other. Each comes with as much overflow of love within himself or herself to give way as he is going to take back in return. It is a true cooperative enterprise. And should the thing be forcefully or otherwise broken, then, if one is in this psychologically healthy state, grief will not be excessive and recovery will be rapid. There are plenty more fish in the sea, and the obvious thing to do will be to be off fishing again.
>
> From what you say of your own love affairs, there does seem to be something a little bit screwy somewhere. I would say that it is fairly minor, though, and is related to the same kind of strict anti-sex taboo in your upbringing as was in mine.

Money concluded by turning to himself:

> You enquire about myself. Of course I had the same background as you, but I have found a slightly different way out. I have had to

> discover that my main technique for escaping from the fear of sex which was inculcated in me was to try and repress the lust factor altogether. This of course worked all right as far as my friendships with girls is concerned, so long as the friendship does not become intimate. Then I begin to panic. That is the key to my love life, and the point at which I have to make the attack. What actually tends to happen, of course, is that I do not find it very difficult to avoid love affairs, so that I have to guard against this and deliberately set out to make them. The picture is additionally complicated in that at primary school I was initiated into sex play with boys, and this tended to make me avoid girls even more, except at purely friendship level. At the present my vocational ambitions occupy such a large part of my life, and I am working so bloody hard in that direction, that my friendships are suffering rather, and I guess it will be like that till the end of the session. The lust factor is more or less adequately discharged through masturbation. I hate wet dreams, the dirty, bloody sticky messes. Do you masturbate, or are you scared by all the old, completely incorrect, horror stories?

I have no idea what his brother thought of this letter, particularly the concluding sentence. And, of course, Money's later life could easily suggest this offhand reference to being "initiated into sex play with boys" indicates denial of his own homosexuality. There is no reason, however, to assume that his brother or anyone else saw "sex play with boys" as anything more than that. It seems likely that the family had become used to this type of letter by this time. Money would continue writing in this format to friends and relatives until the end of his life. The last letter I have seen is from 2001.

One of the central elements of Money's life in Dunedin was his friendship with Janet Frame, author of many novels and books of poetry and arguably New Zealand's most acclaimed writer. They met when she took his class at Otago. He recalled to her biographer, Michael King, his early impressions of her reports on psychological experiments:

> Each one was done as a parable, with birds or other animals acting the role of the investigator, the experimenter and the observer. They were all brilliantly done, and she had the principles one hundred percent correct. Whereas other people I consulted would have failed her for not obeying instructions, I gave her an A-plus for understanding what the experiments were about. (2000b, 65)

Even at this stage, Money was quite convinced that, while "other people" were limited by instructions and rules, he was able to see beyond such restrictions to see true "understanding." Frame's autobiography, *An Angel at My Table* (1984), depicts Money under the pseudonym "John Forrest":

> My escape from teaching was the Psychology class and the Psychology laboratory where we performed a range of interesting experiments and tests supervised by two fresh young lecturers, Peter Prince and John Forrest whom we called Mr Prince and Mr Forrest, but whom I nicknamed H.R.H. and Ash (after Ashley, the fair young man in *Gone with the Wind,* played by Leslie Howard). (59)

She recalled him as fascinating in many ways, not least as a pianist who introduced her to classical music. When she had to hand in an autobiography, the situation became much more complicated than a simple teacher-student relationship:

> I wrote at the end of my autobiography, "Perhaps I should mention a recent attempt at suicide ..." describing what I had done but, to make the attempt more impressive, using the chemical term for aspirin – *acetylsalicylic acid.*
>
> At the end of the class that week, John Forrest said to me, "I enjoyed your autobiography. All the others were so formal and serious but yours was so natural. You have a talent for writing."
>
> I smiled within myself in a superior fashion. Talent for writing, indeed. Writing was going to be my profession!
>
> "Oh I do write," I said. "I had a story in the *Listener* ..."
>
> He was impressed. Everyone had been impressed, saying, "The *Listener*'s hard to get into."
>
> John Forrest looked at me closely. "You must have had trouble swallowing all those aspros?"
>
> "Oh, I drank them with water," I said calmly. (1984, 66)

As a result, Money and two colleagues came to her apartment and convinced her to enter the psychiatric ward of the Dunedin Hospital. When she became more balanced, the hospital elected to send her home, but when she refused to go with her mother, she was admitted to Seacliff, the psychiatric hospital.

Money's memory, as he recalled to King (2000b, 67), suggests he was rather more significant to Frame than her description implies: "I had been the object of her pash since the beginning of the year." Eventually, she was released from the hospital with a diagnosis of schizophrenia:

> My consolation was my "talks" with John Forrest as he was my link with the world I had known, and because I wanted these "talks" to continue, I built up a formidable schizophrenic repertoire: I'd lie on the couch, while the young handsome John Forrest, glistening with newly applied Freud, took note of what I said and did, and suddenly I'd put a glazed look in my eye, as if I were in a dream, and begin to relate a fantasy as if I experienced it as a reality. I'd describe it in detail while John Forrest listened, impressed, serious. Usually I incorporated in the fantasy details of my reading on schizophrenia.
>
> "You are suffering from a loneliness of the inner soul," John said one day. For all his newness and eagerness to practise psychology and his apparent willingness to believe everything I said, his depth of perception about "inner loneliness" was a mark of his special ability. He next made the remark which was to direct my behavior and reason for many years.
>
> "When I think of you," he said, "I think of Van Gogh, of Hugo Wolf ..." (Frame 1984, 82)

Money often asserted his skeptical view of Freudian theory, and his letters suggest he was far from a Freudian, even in his early years. His letters to his family, however, show he certainly was prone to quick, often sexual, interpretations of neurotic behaviour. He clearly extended this to psychopathology in his first encounters with Frame. According to King, Money doubted the diagnosis of schizophrenia, but Frame's creative narratives convinced him, at least for the time. King recorded the full letter in which Money assessed that loneliness:

> "Many people suffer a loneliness of spirit which becomes so overwhelming that at last it forces its possessor to grasp violently at the nearest straw. This is a typical manifestation in our cultural pattern and produces a pattern of behavior which is mostly called love, but which more truly can be called pathological love. It is a search for a redeemer rather than for a cooperator in the mutual enterprise of parenthood, which is the ultimate goal and rationality of true love. When such a person finds a possible redeemer he or she clings possessively for fear of letting such a fortunate looking chance escape."

> This was as close as anybody got at this time to understanding what was wrong with Frame. (2000b, 75)

Money's qualifications and experience were limited, but this diagnosis seems unusually perceptive and mature. While it has psychoanalytic tinges, it is much closer to the lovemap theories that Money would develop many years later. It is typical of Money at his best – articulate, polished, and insightful but also, at the same time, somewhat distant. Furthermore, the views of "our cultural pattern" are at once generally convincing and too sweeping – a combination that would plague Money's work forever.

The letter is rather different from Frame's account of one therapy session:

> A few weeks later I said to John Forrest, "It's awful, I can't tell you, for years I've been guilty about it. It's … it's …"
>
> He waited expectantly.
>
> "It's masturbation, worry over masturbation …"
>
> "It usually is," he said, and began to explain, as our book had explained, how it was "perfectly alright, everyone did it."
>
> The pattern of that "little talk" was so perfect that I imagine (now) a fleeting triumph passing over John Forrest's Freud-intensive face: here was a textbook schizophrenic. (1984, 85)

Frame's writing is often playful, which is perhaps the reason for this note about "a textbook schizophrenic," but the conversation seems to have nothing to do with schizophrenia. What it does depict in miniature, however, as do the letters to his family, is the beginnings of Money's crusade to correct the antisexual world.

All sources credit Money with arranging Frame's first publication. In 1952, once again at Seacliff, Frame was scheduled to have a prefrontal leucotomy, or "lobotomy." She wrote to Money to ask his advice, apparently still viewing him as her source of psychological assessment. He advised against it, but apparently the real reason that one of the twentieth century's major authors was saved from such treatment was the discovery by the hospital superintendent that Frame had won a major award. Still, it is clear that Money was both directly and indirectly responsible for much of the success of Frame's early career. Frame and Money remained close until her death in 2004. Money helped her financially and with accommodation and introductions in the United States. There were vexing moments, but the relationship seems to have been central to both of these world-famous New Zealanders.

One of those difficult times, however, seems to reveal quite a bit about Money. In 1946, Frame wrote the following reply to one of Money's letters:

> There are so many things I wanted to write in answer to your thought-provoking letter, but tonight I can think only of the overwhelming accident that has just happened in my family. My sister Isabel was drowned at Picton last Monday and buried here on Friday. "World is suddener than we fancy it." … I almost cannot bear to be thinking that tonight outside in the dark I have two drowned sisters, even colder than any live people. (King 2000b, 88)

That this should happen after her sister Myrtle's drowning death was overwhelming. King then writes of Money's reply:

> John Money, who had shown inexhaustible patience and sensitivity towards Janet over the previous sixteen months, now made what seemed to be a blunder. He overlooked the fact that, only months earlier, his patient had complained about the "form letters" that well-meaning people had written to the Frames after Myrtle's death. Money himself now wrote what Janet would judge to be another "form letter": "I am deeply grieved to hear of the shocking bereavement you and your family have sustained. Words are so inadequate; I can only offer you the little bit of human friendship and sympathy which is a fragment of help on such occasions. There is also the assurance that time is a kindly physician. Please convey my sentiments to the other members of your family." (2000b, 89)

Frame offers a more abbreviated account in *An Angel at My Table* and then comments, "I remember the complete letter for the shock of its language and my inability to accept the formal conventional expressions of sympathy and to accept that John Forrest was so lacking in imaginative understanding that he could write such a letter" (1984, 94). She then justifies his letter by noting the difficulty of achieving the correct tone in letters of sympathy. King offers the following 1998 comment by Money in a footnote: "A 'form letter' is one that comes off a printing machine and is distributed to many recipients. I wrote a personal letter of which there was only one copy" (2000b, 535).

In the overall biographies of both individuals, this is most definitely a minor event, especially given the permanence of the friendship between Frame and Money, but it still exemplifies an ongoing element in Money's professional life. He never found correction or apologies easy. Once he had done or said something, he seemed to accept the validity of that action or statement, no matter what it was. His consistency was his strength throughout his career, but it was also often a liability. As well, while his sympathy for his patients was usually very evident, his empathy for friends and colleagues was limited. Janet Frame was then somewhat between patient and friend. At this point, Money was moving on, preparing to go to the United States. He of course had no idea how celebrated she would become or how long-standing and important their friendship would be. He undoubtedly recognized his significance to Frame, but at this crucial juncture he did not consider how callous this letter could seem. Fifty years later, he was still unable to acknowledge his mistake and instead emphasized the literal meaning of *form letter.* Even long after, when the admission of an error would have no effect other than to massage someone's feelings, he could not do it. His ability to move intrepidly, even fiercely, in new theoretical and clinical directions was perhaps supported by his recognition that there would be no apologies.

There is one other small element in Money's early involvement with Frame that is not noted by either Frame or King and was perhaps unknown to them. In 1947, Money gave a paper to the British Association for the Advancement of Science Congress in Wellington. The paper, "Basic Concepts in the Study of Personality" (1947), is not unusual, although its emphasis on the importance of "sexual status" and its attack on society's failure to deal directly with sex anticipated Money's future work. A significant part of the paper, however, was devoted to one case study. The patient is not named, but it is clearly Frame. Even in 1947, some of the details, such as the names of her sisters, would have immediately identified her to anyone with cursory knowledge of her biography. Did Money think that no one at such an august conference would know anything about Frame? Or did he think the scientific value of his elucidation of her case overcame any need for confidentiality? He was a young scholar, soon to leave, already thinking about the United States. At the same time, this paper, when linked to that letter, suggests someone who could let professional purpose overtake personal commitment.

It is more than symbolic that Money's journey to the United States was a five-week trip by ship. The trip offered him a leisurely space to transform from bright young New Zealander to beginning American psychologist. He would be a resident at the Western Psychiatric Institute in Pittsburgh. Then, in 1948, he became a doctoral candidate at Harvard's Department of Social Relations. He seems to have had no great difficulty in passing the requirements, but, once again, he conflicted with authority:

> In one written examination, I wrote an essay that to my amazement was not acceptable. I had strayed too far from the party line as taught in the compulsory introductory proseminar. I had written about the universal inevitabilities of being human. To retaliate, I enlarged my essay and it became my second published paper, "Unanimity in the Social Sciences with Reference to Epistemology, Ontology, and Scientific Method." (1991a, 243)

Like his first published paper, "Unanimity in the Social Sciences" is full of sweeping generalizations and not a little hubris. But it was published. Once again, Money felt vindicated in his battle with the small minds of those above him.

Money had to choose a dissertation topic: "The title accepted for the term paper I would write was 'Psychosexual Development in Relation to Homosexuality.' The outline was far enough advanced for me to know that it would not be a term paper in length, but a weighty monograph" (1991a, 244). Then, as a result of a case presented by his professor, Money decided the hefty tome would rather be on hermaphroditism. His first research was at the Massachusetts General Hospital, but he soon went to Johns Hopkins to explore the work being done by pioneering pediatric endocrinologist Lawson Wilkins. Wilkins's stature in the field is suggested by the fact that until 2010 the Pediatric Endocrine Society was known as the Lawson Wilkins Pediatric Endocrine Society. Wilkins offered Money an appointment in 1951:

> During the course of my visit in January, I was interviewed in the Department of Psychiatry, in which I would be academically appointed. The impression I was left with was, on the whole, of a rather lackluster department, its research stranded in the doldrums. Apart from pediatric endocrinology, I would be without spiritual kin in clinical psychoendocrinology, not only at Johns Hopkins but in general. (Money 1991a, 249)

There is an obvious sense of alienation from psychiatry in this comment, an alienation that is multiply important. From 1910 to 1941, Johns Hopkins University (JHU) had been the home of Adolf Meyer, one of the most influential American psychiatrists. In 1930, under Meyer's leadership, JHU had established the first centre for child psychiatry in the world. And yet Money found no home there. Money may have simply been correct about the department being "lackluster," or this could rather be indicative of Money's difficulty in dealing with peers or mentors. Beaglehole and Wilkins were very much exceptions. Yet Money stayed at JHU until he died more than fifty years later. He had a few close colleagues and some excellent assistants and postdoctoral fellows, but even including those exceptions his professional life at Johns Hopkins was "without spiritual kin."

Of course, there could have been kin of a different sort. The last letter to his mother in the Money collection in Dunedin, dated March 5, 1950, refers to his plans for marriage to a writer named Grace Amundsen:

> There are innumerable ways in which it seems that we suit one another admirably. The thing that to me is most important is her complete womanhood, her ability and desire for the feminine role of wife and mother; and her complete mastery of all the arts and techniques of those roles. You will not be able to imagine a more conspicuously happy home life than that which we are both equipped to make; and I think we have a pretty good chance of being good parents. Of sexual compatibility there is no question and that of course is of primary importance.

Given Money's future life and given the complexity of his views of gender and sexuality, it must seem strange that he was apparently so happy with such a limited perspective on "complete womanhood," "home life," and the future of reproduction. The reference to "sexual compatibility" is the one slight note that is typically Money. His sanguine assurance about the forthcoming marital bliss is joined by comments that her stories would be published often enough to provide her part of the income. At this point, Money seems to have had a sexual and economic optimism that would soon change.

The marriage was very brief, and Money made very few references to it. His few friends and colleagues who knew her remember a very difficult person and a consistently vexed relationship. In "Explorations in Human Behavior," he reflected on the implications of the demise of his marriage:

> The greatest forfeiture of this lifestyle is to have had no parenthood. I gave up on that after going through the invincible horror of separating and breaking up a long-term relationship with a woman brilliantly accomplished professionally, sexually, and as a homemaker. Years later, with the advance of psychiatric knowledge, I recognized that a wrong diagnosis of the source of the horror had

> been made. The correct diagnosis was toxic amphetamine psychosis from pills that she had been prescribed for weight loss to control orthostatic albuminuria. (1991a, 254)

It is possibly revealing that Money attributes the end of his one attempt to be in a domestic relationship to a medical misdiagnosis made before "the advance of psychiatric knowledge." Science could have changed all of the psychological torment, likely on both sides, that led to his divorce. There seems to be no suggestion that his next fifty years as a very single sexual adventurer were other than a consequence of an error made before scientific progress could prove the truth. If his wife had received proper medication, he would have become a smiling grandparent enjoying his golden wedding anniversary.

Instead, the rest of Money's life was professionally rich but quite lacking in the elements most would associate with a personal life:

> By trial and error, I discovered the unstressfulness of self-sufficiency in both living and working, as compared with accommodating to supervision or being subordinate. Of course, I have never worked in total isolation but always with assistants and colleagues. Nor have I lived sexually abstinent, but in a give-and-take of sexual visitations and friendly companionships with compatible partners, some women, some men, some briefly, and some with continuity ending only in death. (1991a, 254)

Yet most of those sexual partners seem to have been, in some sense, colleagues. He had sex with men he picked up at cruising spots or in bathhouses, but even his libertine adventures seem to have been primarily experienced with men and women at swinging parties he organized at conferences. He had many friends stay with him in Baltimore, but almost all were either sexologists from other countries or New Zealanders. For the last twenty years of his life, his closest companion and his closest assistant was his niece Sally, who lived next door. Regardless of how famous or controversial he became, his family was always central to him.

It was as if he had decided that he would have only one life – the professional one. This might have been a response to that lack of spiritual kin, to the end of his marriage, or even just to the almost messianic zeal with which he pursued his work, which is visible throughout this book. The possible causes are many. He was always aware of the poverty of his youth and often embarrassed colleagues with his intense frugality. At a buffet at the hospital, he would wander around with plastic shopping bags to gather the leftover food to take home. He made his students and assistants peel off the stamps so envelopes could be reused. Regardless of his income, he thought as a poor person, and he believed that more work would be the only means of ensuring future funding. His commitment to science was immeasurable and absolute. Perhaps this reflects the inferiority that most non-doctors feel in a medical setting, but it also suggests his obsessive belief in the potential of rational inquiry: the truth could be known. Also, he was both extending his New Zealand youth and refuting it. He would be like Janet Frame, someone produced by New Zealand, honoured by New Zealand, and yet judged by New Zealanders as both too superior and too alien to be a proper New Zealander.

In 1956, Money was flying back to the United States and reading a copy of James K. Baxter's *The Fire and the Anvil,* which Baxter had inscribed to him. In the flyleaf, Money wrote,

> I am not an American. I am not a New Zealander, I am an expatriate.
>
> Their foibles and their stamina, neither are hidden from me, and I feel I belong among them. Yet I am an expatriate because I am disoriented. I loved & hated – both. I still love and hate, from which indecision emerges a challenge, the challenge of seeing things that ought to be different, things that I would like to be different.
>
> More deeply, therefore, am I in a quandary, since I am not enthusiastic about being a missionary, not for any creed or cause.
>
> "Perhaps" – I almost hear the words – "you had better go away again, and leave us like we are. The kiwi, after all, doesn't like to be prodded out of the dark."

He later added to this note his happiness at returning to Baltimore: "But if you really want to know, I went back to the anonymity of a large city."

He might have felt he was no longer a kiwi, but, to paraphrase Graham Greene, New Zealand made him. After New Zealand, his personal life would be anonymous, except for the parts that connected to New Zealand. Otherwise, his life would be professional. There would be no lovemaps, except for the ones he wrote of in his books.

CHAPTER TWO

Fixing Sex

Intersex

As Anke Ehrhardt (2007) has stated, Money created the field of pediatric psychoendocrinology. He also created our contemporary tendency to see intersex as a source of understanding all aspects of gender and sexuality. Humans have always recognized the existence of someone who combines both sexes. Presumably, Hermaphroditus appears in Greek mythology to represent a spiritual reason for this blending. Until recently, the hermaphrodite was seen by most, throughout the world, as either a monstrous mistake of nature, suitable for display in circus sideshows, or else a mystical purveyor of shamanic truths not available to those who are clearly of either sex. Money saw something quite different. In 1955, he published "An Examination of Some Basic Sexual Concepts: The Evidence of Human Hermaphroditism" (1955b). The most noteworthy element of the article is probably the claim that gender identity is more connected to early "life experience" than to chromosomal sex or gonad sex. Yet he also included the observation that sexual orientation seems "undifferentiated at birth" (308) and is similarly a response to experience. In other words, he found that, regardless of biological disruption, intersex individuals were "normal" in behaviour and could be treated as "normal" by society.

Money composed a protocol based on the treatment at Johns Hopkins that he believed responded to those observations. The protocol suggested that any indeterminacy of sex, as in the appearance of the genitals, should be corrected early so a child could be brought up in a gender without any doubt in the minds of parents or anyone else who participated in his or her rearing (1955c). Money believed this would do much to establish resolutely that category of "normal" for both the individual and those who interacted with the individual. Thus, "life experience" would lead to a clear male or female identity. Many have since seen surgical "correction" as unnecessary medical intervention, especially given there are various conditions that offer no absolute direction as to which is the best choice of gender. Once Money established his protocol, he was consistent – to a fault. The protocol could be called *doctrinaire,* a word that would have offended him greatly. His theories were highly developed and explicit, but he intended them to be open-handed and elastic, with plenty of room in their largeness for change. He always argued that science should be the absolute arbiter but that science required correction. Yet this intention of change and correction came with Money's tendency to see his own opinions and theories as inevitably true.

The *OED* states that *intersex* refers to "an abnormal form or individual having characteristics of both sexes." This definition obviously applies to all of us, as we all have "characteristics of both sexes." How then could it be "abnormal"? The term must be narrowed. The usual way it is narrowed is by limiting it to those who have genitals that are visibly different from the norm and/or who have a chromosomal difference from the norm. Even this does not include all the people who are considered in medical science to be intersex. Thus, even more than many others, the term *intersex* is a catch-all for anyone that medicine wishes to include. Similarly, but somewhat in opposition, those who call themselves intersex tend to be those who believe themselves mistreated by medical intervention that took place because they were deemed intersex and thus needing treatment, usually surgical.

The full title of Katrina Karkazis's book is *Fixing Sex: Intersex, Medical Authority, and Lived Experience* (2008). When I first heard the title, I was envious of how much those first two words resonate. Medical science assumes that an "abnormal" human has a problem that needs to be "fixed," repaired. But there is more. This abnormality, to use the favourite word of the medical literature on intersex, is "ambiguous." In other words, it needs to be "fixed" or stabilized. The word *sex* might seem obvious – either male or female. But this binary, of course, leads to Money's word, *gender.* Regardless of our faith in biology, we cannot understand the sex of a human without understanding the gender, whether the person is masculine or feminine. Then there is the word *sex* as an activity, as in "having sex." This might seem irrelevant to establishing the biological sex of a human, but, as all the literature shows, sexuality is central to all thoughts about intersex.

Anyone seeking an understanding of the issues associated with intersex should not look to this brief chapter. There are now many books on intersex, most notably *Hermaphrodites and the Medical Invention of Sex* (1998) by Alice Domurat Dreger; *Lessons from the Intersexed* (1998) by Suzanne J. Kessler; *Intersex and Identity: The Contested Self* (2003) by Sharon E. Preves; and *Bodies in Doubt: An American History of Intersex* (2009) by Elizabeth Reis. But after exploring the literature widely, I have yet to see a text that equals *Fixing Sex*. It maintains a dispassionate fascination and yet comes so close to that impossible ideal of objectivity with a full portrait of the field. The present chapter seeks only to offer a basic understanding of Money's involvement in intersex treatment and some of the central issues around that involvement. Not that Money's work is a minor aspect of the field. As Karkazis notes,

> By the 1970s, medical protocol almost exclusively reflected Money's paradigm; the model's tenacity results in part from its institutional embeddedness. Money's paradigm formed among clinicians "a consensus rarely encountered in science." (2008, 62)

The quotation in Karkazis is from Kessler's *Lessons from the Intersexed* (1998), which comes to the same conclusion. This consensus on the consensus suggests that "Money's paradigm" was the rule for almost everyone.

Money's theories began with his PhD dissertation, "Hermaphroditism: An Inquiry into the Nature of a Human Paradox" (1952). The base of the study was an examination of 248 cases in the published literature, but Money was often frustrated by the lack of psychosocial information offered in that literature. He added to those examples by performing in-depth personal interviews with ten individuals who had been referred to him by endocrinologists and urologists in the northeastern United States. The dissertation devotes two hundred pages to these ten cases, providing sensitive, nonjudgmental biographies of quite remarkable breadth and depth. Even at this stage, Money had already developed his talents as a sympathetic and precise interviewer, a characteristic that both patients and colleagues noted through the rest of his career.

Money used the conclusion to his dissertation as his abstract, which might be seen as his way of emphasizing what he saw as the most important information to be gleaned from the dissertation. He concluded that, while the ten detailed case studies offer much more knowledge, they do not at all contradict the assumptions he made from examining the published material. His most striking discovery was simply that the intersex individuals were like everyone else:

> The findings are somewhat disconcerting, for one would not have been surprised had the paradox of hermaphroditism been a fertile source of psychosis and neurosis. The evidence, however, shows that the incidence of the so-called functional psychoses in the most ambisexual of the hermaphrodites – those who could not help but be aware that they were sexually equivocal – was extraordinarily low. The incidence of neurotic psychopathology of the classic types, sufficiently severe and incapacitating to be unmistakable, was also conspicuously low. (1952, 6)

Thus, it seems that whatever sex was assigned to the individual proved to be acceptable, in most cases. Money's conclusion is best summed up in an article he wrote a few years later, with his colleagues the Hampsons:

> The sex of assignment and rearing is consistently and conspicuously a more reliable prognosticator of a hermaphrodite's gender role and orientation than is the chromosomal sex, the gonadal sex, the hormonal sex, the accessory internal reproductive morphology, or the ambiguous morphology of the external genitalia. (1957, 333)

As a first study of the psychology of intersex individuals, this suggestion must seem quite amazing, but actually others had come to similar conclusions. The most significant source for Money's published cases was a 1937 text called *Genital Abnormalities, Hermaphroditism and Related Adrenal Diseases* by Hugh Hampton Young, professor of urology at Johns Hopkins. Young (1937, 199) observed how easy so many of the patients were with this unusual condition and stated, "It is interesting to note that the libido or emotions of the patient corresponded to the manner in which the individual had been reared in all cases but one (No. 11) in which no note was made although a gonad of the opposite sex was also present." Albert Ellis, another sexologist with a controversial reputation, came to similar conclusions in a survey of case studies he made in 1945. He presented a series of rejections of any suggestion that chromosomes are destiny: (1) "This seems to indicate that physiological factors are not decisive in determining the masculinity or femininity of pseudohermaphrodites" (118); (2) "This would seem to indicate that biological factors are not decisive in determining the masculinity or femininity of true hermaphrodites" (119); and (3) "The inference is again clear that biological factors in themselves have a relatively slight influence in the causation of masculinity and femininity" (119).

Thus, two surveys of the literature agreed with Money: all found a surprising normalcy among hermaphrodites, a normalcy that primarily reflected the gender of upbringing. But Karkazis (2008, 64) writes, "In this context, the domination of Money's treatment paradigm for nearly four decades appears like an aberrant period of a largely uncontested viewpoint nestled between a century of debates prior to its formulation and the past twenty years of modern debates urging its revision and even its demise." These two studies, sensitive to the psychology of the intersex individuals, seem to have come to quite similar conclusions. There were other views, but those views were from analysts much less significant than Young and Ellis. The primary "context" to which Karkazis refers, however, is the medical treatment of intersex newborns. There no unanimity appeared until Money's protocol seemed to resolve the confusion and provide an answer to a deceptively simple question: when is a child in need of surgical intervention?

In some ways, this question seems surprising after Young and Ellis. If most intersex individuals find it possible to be normal in whatever their assigned sex, why intervene at all? Just name each child male or female and be done with it. The history of intersex, however, particularly as traced by Reis (2009), shows constant confusion in medicine and the law as to how to treat intersex individuals. It is as though the potential normalcy of the intersex individual, so clear in Young, Ellis, and Money, was just too surprising and counterintuitive to be accepted. Perhaps this was the primary reason for the domination of Money's treatment paradigm. The intersex individual was normal and could be treated as such, but society could not accept this was the case. The individual needed to have the biology regularized in some way so that society could see the normal in order to begin to accept it, and Money's protocol offered this opportunity. Just saying "boy" or "girl" could not be enough.

Some intersex activists have gone so far as to suggest that even the declaration of gender is not necessary. The majority believe that an assignment of gender to each newborn is inevitable, but it might be possible to create an atmosphere of indeterminacy so that an individual can make decisions later on about what sex to assume. Activist and scholar Morgan Holmes (2008, 154) asserts that "the only constant in the management of intersexuality and the assignment of sex to infants is the desire to match male chromosomes with social meaning." She continues, "The categorization of intersex is related to ideological commitments to a presumed binary 'nature' of male and female, coupled with a paradoxical assumption that gender is so fluid that we are entitled to make of infants and children whatever we will" (155). According to Money, however, it is not a "presumed binary 'nature' of male and female" but rather an assumption that society is organized according to a "presumed binary 'nature' of male and female" that is at stake in the categorization of intersex. Money himself provided an entertaining analogy when he made a trip north after attending a conference in Sweden and encountered the midnight sun:

> I had arranged that boat ride so that I could say, "I have heard about it, I have read about it, and I have intellectually accepted it, but now I have seen it with my own eyes: day and night are not a universal feature of the cosmic order. I can bear witness to the demolition of a once secure belief in the absolute difference between day and night."
>
> The special significance of this experience for me lay in its similarity to what had already happened to me regarding the difference between male and female. That difference had also been implanted in my mind since childhood as an absolute and eternal verity which, in adulthood, had been progressively demolished by my studies in hermaphroditism. (1993a, 148)

Money (1978c) suggested that these assumptions are so "absolute and eternal" as to be almost unshakeable:

> We do feel our gender identity so intensely that any threat to our work, legal status, games or fashions as sexually dimorphic seems to be taken as a threat to the very well-being of our sex organs. The true sex roles, the authentic, bona fide sex-erotic differences have to do with the reproductive function itself, and they carry through to the having of a baby and the care of it. Those are the basic sex differences. In my mind I sum them up in saying that the only imperative sex differences that are irreducible are that men impregnate and women menstruate, gestate and lactate. All else is cultural, historical, and relative.

Money (1975b, 48) believed that our assumptions about sex differences are to some degree a product of our desire to see everything in binaries: "From our

Platonic and Biblical forebears, you and I have the heritage of dichotomizing. Dichotomies are dangerous, for they may be wrong." Though they may be wrong, they are the way we think. All of Money's theories were based on his belief in the inevitability of binaries and the likelihood of their error: they are always our assumption, and they are often wrong.

Money's belief in the inevitability of the gender binary convinced him of the necessity of coming to an early decision about the gender of an intersex child, particularly one that resolves the confusion of the parents: "Thus they can transfer their confidence eventually to the child, as well as conveying it, in discussion, to others in the community" (1969e, 213). Money hoped to enlist "the community as doctor," and this cannot be done by attempting either secrecy or an assertion that a child is somehow beyond the label of male or female: "The basic sexual dichotomy is traditional to the community in dealing with a person in certain socially-defined ways. If ambiguity is not resolved, a child is assigned a special status, typically as a freak" (213). The answer is "a rhetoric of normality," in which some reference to illness or disability "must fit within the patient's sense of what is normal" (214). Thus, regardless of the success of so many intersex individuals in adapting to whatever gender they had been assigned, Money saw improvements in medical science as (1) enabling a much better projection as to what the most appropriate gender should be and (2) enabling much better surgery and hormonal treatment to make that appropriate gender as normal as possible, in both presentation to the outside world and psychological well-being for the individual. The choice of gender might be in some sense an error, but it would be a larger error to expect the individual to function in an ambiguous gender.

This emphatic choice of one gender reflects Money's theories of gender role and of the complementarity of gender identities. Money provided this definition in 1955:

> By the term, *gender role,* we mean all those things that a person says or does to disclose himself or herself as having the status of boy or man, girl or woman, respectively. It includes, but is not restricted to sexuality in the sense of eroticism. Gender role is appraised in relation to the following: general mannerisms, deportment and demeanor; play preferences and recreational interests; spontaneous topics of talk in unprompted conversation and casual comment; content of dreams, daydreams and fantasies; replies to oblique inquiries and projective tests; evidence of erotic practices and, finally, the person's own replies to direct inquiry. (1955b, 302)

To the assumption of the dichotomy between man and woman, Money added the theory of complementarity, that each person must identify the roles of both man and woman in order to live in the one and recognize the difference from the other. Thus, for Money, the basis of gender experience is a binary, one that perhaps very few people can straddle but that society and the vast majority of its inhabitants find absolute. You must be one or the other, but how do you decide? In a 1986 interview for *Omni* magazine, Money assessed the situation as follows:

> I would say that you still have two camps, so that there's still a lot of people in medicine who are looking for an absolute criterion, and they look for it in the genes, and maybe they consider the gonads a bit. And then they'll force the child to grow up as a boy without a repairable penis. But there's fortunately an increasingly large number who see that one of the very important criteria in making a decision for a newborn baby is whether you can effect surgical repairs and plan a rehabilitation for that baby's grown-up life, with the sex-life included. (1986a, 61)

For many intersex activists, the question of whether one is male or female is irrelevant. Surgery should never be performed on an infant unless there is a clear case of medical necessity, as when there is a likelihood of testicular cancer. Money, however, saw a different necessity, in which all involved in caring for the child must be convinced that the child can inhabit a gender normally. This began with the parents. All commentators on intersex note the immediate question after birth: "Boy or girl?" In early childhood, almost all children are profoundly gendered, with boy parties and girl parties, with children identifying even the most unisex object as a "girl thing" or a "boy thing." Even those who believe we should avoid dividing all the world into pink or blue recognize that the binary is very difficult to escape. Most parents are unable to try to present a unisex child to the world, so a decision must be made. The obvious question is whether surgical intervention is required to make that decision. At the beginning of the twenty-first century, the decision is not as simply chromosome-centred as Money feared, but, in most cases, the decision is a combination of chromosome, genital appearance, and the likely hormonal future of the child. It is impossible to predict in all cases, but some conditions that at birth seem female can, for example, offer a projection of virilization at puberty. A fictional representation of this process provides the plot for Jeffrey Eugenides's novel *Middlesex* (2003).

Some of the conditions seem to contradict directly simple biological assumptions. The most classic case, one very often noted in the popular press, is complete androgen insensitivity syndrome (CAIS). This person has the XY chromosomes of a male, but failure to respond to hormones at any time in development leaves her as a fully functioning woman, with the exception of reproduction. While different CAIS cases might require some surgical treatment, there can be no doubt about the gender assignment: a CAIS person needs to be considered by all and sundry as a female. Such a resolute response cannot be given for all intersex cases, but a sex can be assigned with a lesser or greater hope of continuity depending on the condition. If the child later decides on sex reassignment, one might consider it at least somewhat similar to the choices a transsexual makes, albeit with a very different history. In other words, a person in one gender role decides to be in another gender role because of a felt gender identity that is different from the assumed gender identity.

The question of genitals, however, is never absent from the discussion. Money was clear from the beginning of his work that a viable penis is necessary for a male gender role and that an "enlarged clitoris" would inhibit a female gender role. Thus, in 1957 he wrote, "Though the sex of rearing could transcend external genital morphology in psychologic importance, absence or correction of ambiguous genital appearance was psychologically beneficial" (336). He extended this need for correction of ambiguity in 1969 when he observed how social stigma can be avoided: "The less visible the signs of illness or defect, the more likely will a normalized version of the case be accepted" (1969e, 214). In 1966, he wrote, "No one should be assigned to a sexual status when the sexual morphology, however corrected, will forbid sexual performance in that status" (1966c, 447). Presumably, Money completely rejected the idea that a man without a long penis could accomplish anything that could be called "sexual performance." He said in an interview, "I can assure you from my experience with patients that to have to negotiate life in this society as a man without a penis is just too difficult. The life outcome is almost always tragic, and suicidal depression is not unknown" (1977g, 26). Earlier he had offered a more complex view:

> All discussion of psychosexual orientation in hermaphrodites is contingent on a simple fact of social psychology and of anatomy: when a baby is born someone looks at the genitals and announces their gender. There is, after all, no other way in a delivery room of knowing a baby's sex! The common sense of this custom means that, more

> often than not, assigned sex agrees with external genital morphology in those hermaphrodites who later are investigated psychologically. This agreement between assigned sex and external genitalia is more than a matter of the common sense of custom. For the person with ambiguous-looking genitalia it provides two assurances. First, it makes surgical repair of the genitals practical so that the assigned sex can be even more definitely asserted. Thereby the parents are enabled to rear their child secure in the conviction that they are rearing either a boy or a girl and not some freakish half-boy, half-girl whose gender they may one day be obliged to change. Unanimous evidence from rearing and bodily appearance is the best guarantee that the child will grow up also secure in the conviction that he is a boy, or that she is a girl, unbeleaguered by paradox and doubt. Second, agreement between assigned sex and external genitalia ensures for the hermaphrodite that there would be no contradiction between the sex of rearing, on one hand, and genital practices and location of genital sensation, on the other. (1954, n. pag.)

The methodical rhetoric of this argument is exemplary. As a teacher, I feel like using it as an example of clarity for my students. No doubt this clarity on a topic that confused many doctors and all parents was no small contribution to the ensuing power of the Money protocol. As so often in Money's work, language had significant power. Many would reply, too much power.

Much has changed today, such as improvements in phalloplasty, the creation of a prosthetic penis. The result does not function the same way as a penis, but it is a reasonable approximation, something that could not have been said when Money wrote this. Perhaps more important, trans men – men who were born in bodies that medical science believes to be incontrovertibly female in sex – have done much to change the assumptions about men and penises. I first encountered a photo of a nude trans man in the mid-1990s and was quite flabbergasted at this image of someone so clearly male without a penis. I doubt I, or anyone with a reasonable degree of sensitivity to the transgendered world, would have a similar reaction today. We have come to accept the man without a penis as a distinct possibility. This is a great change from when Money and six colleagues published "Micropenis. 1. Criteria, Etiologies and Classification" (1980d). They went into great detail about all aspects of the condition but never provided the slightest suggestion that a person could be a man without a penis. Money seems to have continued to believe this to be the case until his death.

Money thought that an enlarged clitoris was similarly forbidding of sexual performance. Money's own dissertation had identified women who had very large clitorises that they found sexually satisfying, and yet in his own work he invariably counselled reduction. Many have suggested that this reveals Money's lack of concern for women's sexual response as such reduction can inhibit response (Minto et al. 2003). There are many examples, however, in which Money asserted the need for women's sexual liberation, in which he most definitely included sexual satisfaction. In terms of the specifics of sexual response, moreover, Money made his views clear in a 1974 review of Seymour Fisher's book *The Female Orgasm*. Money noted Fisher's error in assuming "there is no relationship between psychological health and orgasm." Money concluded that the book "is useless for one looking for answers, unless one is satisfied with a conceptualization of the female orgasm as a dissociated psychological occurrence somewhat less important than the hiccoughs" (1974a, 400). Thus, Money's belief in the need for clitoris reduction was not a dismissal of the importance of the female orgasm; rather, it was an assumption of the ultimate importance of genital appearance in a functional gender identity. He hoped for continued full sexual response, but his first concern was the normal genital appearance that he believed to be required in a comfortable, coherent gender identity.

Whether Money was correct, his attitude reflected the views of his time, and this attitude was based not on what he thought of gender roles but on what he thought was the best response to society's thoughts on gender roles. Still, it seems quite amazing to me that the treatment of intersex children is seen as simply a product of Money's theories. It is not difficult to see why his ideas were persuasive in the treatment of intersex, but they remained the opinions of a psychologist. The urological surgeon makes the decisions, regardless of the value or fame of any psychological theories. In *The Adam Principle* (1993), Money made it clear that the procedure was not new:

> Living as a male with a micropenis is as difficult as living as a male whose penis has been accidentally amputated. That difficulty is circumvented if the pediatric surgical and clinical care of a newborn micropenis baby is programmed to allow the child to grow up to be as complete a female as possible. This was the policy adopted as early as the 1940s by Lawson Wilkins, the pediatric endocrinologist with whom, in 1951, I came to work at Johns Hopkins. It was matched by the same policy applied to cases of male hermaphroditism in which the deformed penile organ was excessively minute. (1993a, 167)

Money was identified as the progenitor of the procedure, although he did not begin it, even at Johns Hopkins. The assumption that a male without a penis should become a female was established well before he arrived. But then again, anything in which Money was involved tended to become centred on him eventually. A good example is the careful and considered study *Sexing the Body: Gender Politics and the Construction of Sexuality* (2000) by Anne Fausto-Sterling. This book has become a central text for all discussion of gender and intersex, probably the Money and Ehrhardt of today. The index does not even have a category for *urology,* but Money receives a lot of space. In the sections on both intersex and other gender issues, his theories are often presented as influencing a problematically insensitive response to those who don't fit his paradigms. For many reasons, "Money" became the label for the standard treatment of intersex.

The first explanation would be historical – how what could have been "Hampson, Hampson, and Money" became "Money." This psychologist who fell into the study of hermaphroditism was then brought to Johns Hopkins by Wilkins to assist two medical doctors, the Hampsons. That relationship rapidly deteriorated. According to Money, Joan Hampson was all right, but her husband, John, was lazy. Richard Green, Money's first student, later a colleague and an eminent sexologist himself, notes that by the time he arrived at Johns Hopkins John Hampson and Money no longer spoke (2008, 611). Money included his own account of his experience with the Hampsons in his papers, a confrontation that makes it clear that, in his eyes, Money did all the work on anything that was published under their three names. In his collection of papers, every copy of their articles includes a pencilled note, "Written by John Money." The Hampsons soon left the field of gender studies, and thus Money became the sole source of the published work on intersex, including the definition of gender, emanating from Johns Hopkins.

As well as being known as the source of many publications on intersex, Money made constant claims to be an originator, as enshrined in the title of his book *A First Person History of Pediatric Psychoendocrinology.* The emphasis should be on that "first" as Money asserted that "much of the early history of the field is also my personalized scholarly and research history" (2002a, 1). He used this work as the base to become, as Green refers to him in the preface, "Dean of American Sexology" (2002a, vii). While a number of his books have little to do with intersex or endocrinology, they were all presented as emanating from an expert, and the definition of his expertise was his work in the field of intersex, where he was a permanent presence. He referred many times to the importance of his "longest continuous NIH [National Institutes of Health] grant" for enabling his longitudinal studies of intersex. In 1999, he wrote one of his many letters confronting what he thought to be the short shrift other researchers gave to his work to the *Journal of Urology.* Editor-in-chief and urologist K.I. Glassberg wrote a reply:

> Money should be acknowledged for his lifelong effort in evaluating the longitudinal outcomes of intersexual individuals. He should also be acknowledged for pioneer studies on the significance of patient age at the time of gender assignment or reassignment in ultimate gender identity. Unfairly, some have accused Money of ignoring "nature" in the "nature versus nurture" debate. However, that never was the case and he always has acknowledged the contributions of nature and nurture.
>
> Money has almost been a lone ranger in his efforts at long-term follow-up. Yes, there is the occasional report on an intersexual individual assigned a gender who undergoes gender reassignment later in life, such as the "John-Joan" case. We learn from these isolated cases but not as much as we learn from the longitudinal studies of Money. (2000, 926)

While Glassberg seemed to erroneously call "John-Joan," or the Reimer case, "an intersexual individual," the important comment is that the many publications by various people on that one case are exceptional, unlike Money's constant publications on many aspects of intersex. Depicting Money as "a lone ranger" is no doubt too true, in both positive and negative ways. To continue the analogy, it is easy for a sniper to pick off a lone ranger, much more so than the leader of a great army. Yet to Glassberg, Money was not only an interesting theorist; he was the source of most of the follow-up material on intersex beyond the anecdotal. When the Reimer case received so much media attention, the general public quickly forgot the broader picture noted by Glassberg.

Money also came to the fore because he was an extensive *self-actualizer,* to use a term he no doubt would have hated for its lack of scientific rigour. But he invariably turned one study into many publications or one thought into many. One of the stranger moments of this might be the brief period during the early 1960s when he became very interested in dyslexia. This interest arose out of the endocrinological clinic, through the connection between reading disability and the intersex condition Turner syndrome, a cytogenetic disorder (Money 1965e). From a brief conversation that Money had with a colleague in 1960 came a symposium on dyslexia, which led to a book on the same topic he edited in 1962. That year, in October, the *Baltimore Sun* published a story on the five-storey research clinic on dyslexia that Money would head, and Money's papers include letters to a variety of specialists at Johns Hopkins University Hospital whom he hoped would be part of the clinic. The idea would soon end, the victim of a lack of funding, but it represents how quickly Money's plans could grow, especially early in his career. The move to dyslexia was representative: as the most famous name in intersex studies, he had power in many directions. It was no doubt an overstatement to refer constantly to "the Money protocol," but it served Money's own purpose in the wide-ranging treatment of intersex patients but also as proof of a range of other theories, including theories on sexual orientation.

Albert Ellis's 1945 study claimed that its primary purpose was to help the intersex:

> [This study is] a plea to the members of the medical profession, to whose lot the firsthand study of hermaphrodites invariably falls, to consider it hereafter their scientific duty to see that thoroughgoing psychiatric and psychological investigations, in addition to the usual physiological ones, are made of all hermaphroditic individuals who may come to their attention. (108)

Ellis went on to state, apparently obliquely, that "regarding the matter of homosexuality, there have been, and are still, two opposed viewpoints concerning its origins" (108). Many might question why this should be a central psychological concern for a paper on treating intersex, but Ellis noted that "in the case of hermaphrodites we often have a beautiful experimental situation all set up for us; and all that we need is to throw added light on the question of normal and abnormal sexual behavior is to observe the sexual psychology they display" (108).

The combination of a clinical and a research interest in intersex might seem like a standard response to any unusual condition. The difference here, however, is that there are multiple research interests based not on the needs of the condition but on that "beautiful experimental situation all set up for us." To look at it from a slightly different position, Morgan Holmes (2008, 13) asserts that intersex is "overburdened with signification."

Like Ellis's study, Money's dissertation used intersex for broad-based claims about sexuality:

> The findings of this review of 248 cases of hermaphroditism carry implications for the two important aspects of psychological theory. The first has to do with the origins and determinants of libidinal inclination, sexual outlook and sexual behavior. In the face of the evidence it appears that the presence or lack of libido is clearly a function of the presence of sex hormones, regardless of their biochemical structure or their source of origin. The evidence weighs heavily, however, against the conception that individual erotic preferences – the direction and goal toward which libido is exercised – bear a direct or precise relationship to unlearned determinants. It does not appear feasible to ascribe these aspects of libido to a basis which is commonly described as constitutional or instinctive, organic or innate, unless it be specifically in terms of the localization of erotic sensation in the genitalia. The evidence weighs even more heavily against the conception that the more general aspects of sexual outlook and sexual behavior – in contrast to the specifically erotic aspects – bear a direct or precise relationship to unlearned determinants. In brief, it appears that psychosexual orientation bears a very strong relationship to teaching and the lessons of experience and should be conceived as a psychological phenomenon. (1952, 5)

Thus, Money used the intersex study to claim that, while sexual libido might be innate, the sexual object is not. At this point, very early in his career, he was both defining the importance of the libido and suggesting that the specificity of the sexual object might not have the import many were claiming. This object is treated by most people, regardless of sexual orientation, as central to sexual expression and is, throughout society, central to gender role. The ubiquitous question "Is he gay or straight?" has nothing directly to do with the "erotic sensation in the genitalia." Money was writing about intersex, but he also seemed to be making a very broad claim that not unlearned but learned determinants are the primary control of psychosexual orientation. There is something from experience, probably somehow connected to that erotic sensation, although perhaps only circumstantially connected, that leads to each person making a sexual object of someone of the same sex or someone of the other sex – the one other sex.

But Money went even further than these questions of gender to the most general assumptions of the sources of psychological well-being and neurosis:

> It is not inconceivable that hermaphrodites are less likely than normal individuals to develop symptoms of psychopathology, because their sexual conflicts – and the case histories leave no doubt that they had them – are easier to cope with, being more tangible. Such an hypothesis does not stand up very well under the weight of evidence from the cases studied in psychological detail, namely, that hermaphrodites appear to have symbolic psychosexual fantasies similar to, and as intangible as those usually found in, anatomically normal people reared in the same sex as they, in addition to their special hermaphroditic problem. Apparently, therefore, sexual conflicts and problems are not in themselves sufficient to induce psychosis or neurosis. This inference is by no means untenable in the light of the evidence afforded by the low incidence of nonorganic psychoses in hermaphrodites. Further, in view of the infrequent incidence of neurosis among the entire group of vulnerable individuals, it appears necessary to postulate that sexual problems and conflicts in themselves are not sufficient to induce neurosis, unless at least one other variable is present. The evidence suggested that in some instances the additional variable may pertain to somatic phenomena. In other instances, notably those of change of sex imposed during childhood, the evidence pointed to an additional variable which was exclusively psychological in nature and bore a close relationship to teaching and the lessons of experience, and in all probability to age. A study of the detailed reports in Part Two throws provocative highlight on this elusive variable. On the whole, however, one thing is clear: traditional and contemporary theories which ascribe the origins of nonorganic psychosis and neurosis exclusively to psychosexual conflicts and problems must be suspect. They require re-examination, if not modification and revision. (1952, 7)

So much for Freud. But also so much for seeing intersex individuals as psychologically damaged. Their unusual gender history might be thought to make them less prone to psychopathology because they have spent lives recognizing the pitfalls of gender and sexuality, but instead they seem to have the same psychosexual confusion that we all do. On the other hand, perhaps they should be more prone to psychopathology because of the deleterious effects of indeterminacy and even more psychosexual confusion than that endured by the rest of us. This also does not seem to be the case. They seem, to use that very loaded term, *normal.*

The history of Money's work on intersex presents a quite classic example of the ambivalent relationship between the clinician and the theorist, with a third character – the researcher – caught in the middle. In *As Nature Made Him,* John Colapinto describes two intersex patients whom Money treated. "Charlie Gordon" is described as very upset because of publication of intimate details of his life in ways that he thought were insufficiently anonymous. Given Money's early treatment of Janet Frame, in which he gave a public lecture on her case that only slightly hid her identity, this seems quite possible. While Money seems to have had scrupulous intentions to maintain the anonymity of his subjects, his methods were not always similarly scrupulous. In part of his attempt to liberate American society's views of sex and gender, he often asked patients to appear before medical students, and Charlie Gordon was one. The other example in the book, "Paula," seems to have had more of a purely clinical experience and was much more positive about Money.

The larger point that Colapinto makes about intersex is that sex reassignment of infants has become rejected by society as a whole. And yet any attention to work done in contemporary hospitals by urologists shows that it continues to happen. If sex reassignment means treating a chromosomal male as a female, then it is inevitable in the case of CAIS. Sex reassignment happens in many cases other than CAIS when a urologist deems it in the best interests of an infant. Still, there is no question that the general populace does not accept the idea that sex reassignment can be a medical necessity. They see sex reassignment as an arbitrary move that interferes with what Money himself recognized to be possibly the most profound binary in human understanding – gender. He considered it to be up there with good and evil and nature and nurture. Regardless of the impossibility of maintaining any of these binaries as absolute, society wants them to be absolute, and sex reassignment seems to offend that.

Some intersex activists have a rather different concern. Many wish to reject these binaries as much as possible. If an individual is neither male nor female, so be it. One partial interpretation of this rejection is the desire that most humans have to be accepted as what could be called normal, not in the statistical sense but in the sense of being acceptable. So if you are neither male nor female, that must be normal. Most of us have our beginning selves accepted as normal. This idea of the original normal self contributes to the recent move wherein many state and private insurers are refusing to fund male circumcision, but it is also an element of various religious prohibitions against body modification, such as tattoos. Such cultural rules often respond to a belief in the body given by God that must be presented in the same form at the end. As we entered the world, so should we leave it. Most intersex activists are not pursuing a specific religious dogma. They are, instead, following the general human concern that surgical intervention should never happen except when absolutely necessary to sustain life or when the patient elects to have it. An infant patient is not able to make that decision. Sex reassignment is one more invasion of the individual. They reject surgery for the same reason that transsexuals seek it – to be the persons they consider themselves to be.

Thus, Colapinto and others do not want to accept stories from those such as "Paula" who were happy with their gender reassignments. Nor does Colapinto wish to see Money for what he seems to have been – a person with a profound interest in his patients. Another of his patients, whom I will call "Bob," refused to speak to Colapinto once the latter's agenda became clear to him and later wrote to Colapinto that "I hope you rot in hell." Bob often appeared with Money in front of medical students and stated that it was only Money's emotional support that enabled him to do so. He thought that, while the surgeons at Johns Hopkins treated him "like a piece of meat," Money was consistently supportive. Money even gave Bob his home phone number. Over the years, Bob would often phone Money with concerns, a couple of times in the middle of the night. He remembered Money as a better parent than his own proved to be. After the publication of *As Nature Made Him*, Bob became a bit of a pro-Money activist. He asserted that the many patients he contacted had no recollection of the sexually invasive and insensitive Money that Colapinto depicts. Bob asked them to speak publicly, but nothing came of it.

This appreciation might seem somewhat surprising in that Money was bluntly direct with his patients in his pursuit of his agenda of liberation. He asked his intersex patients many questions that seem to have had little to do with their condition and had much more to do with helping him to develop a larger portrait of the possibilities of sexual diversities. Just as he had used his intersex data years before to make anti-Freudian claims, so he came to do the same in his fight against sexual repression. In this passage from *The Adam Principle*, he used intersex to confront what he saw as false religion, false medicine, false science, and fear of openness about sex:

> People who have come under the influence of the tyranny of the chromosomes and the testicles, whether they be the baby's parents or

> their relatives and friends, cannot relinquish the idea that, if they raise the baby as a girl, the real sex is male and the girl is really a boy, and a girl only by clandestine conspiracy and collusion against nature. Here are the ingredients of the unspeakable secret of not being what you think you are. Stigmatization is conveyed both by not speaking the unspeakable, or by speaking it in a cruel and traumatizing way.
>
> Appeasing the unspeakable monster of sex by sacrificing some victims of birth defects of the sex organs on the altar of stigmatization is part of the price exacted by our taboo-ridden, antisexual culture. Those who escape vulnerability to stigmatization do so on a more-or-less hit and miss basis. (1993a, 164)

Colapinto ([2000] 2006, 36) writes what might be seen as an offhand interjection, in parentheses, to assert that "Money himself was a psychologist and did not possess a medical degree of any kind." The "of any kind" suggests how emphatically Colapinto wishes to make his point. But one might suspect that Money himself recognized this exclusion. For him, the false science of the simple truth of chromosomes and testicles represented what doctors of medicine could control. There is another tyranny here, moreover, the tyranny of a "real sex" that is male. Money rejected masculinist assumptions that he thought led to dysfunctional treatment of women and men. Similarly, however, he believed anti-male assumptions to be anti-sexual, leading to prudish repression. For him, the belief in the necessity of a viable penis was a reaction to both masculinist and anti-male assumptions. He would avoid letting babies be required to perform as men just because they had male chromosomes if he thought a successful male identity was not possible. For him, a male identity required male sexuality – not necessarily heterosexuality – and a male sexuality required a penis. And the penis must be there from birth. In a 1965 article, he treated the inadequacy of phalloplasty as almost a natural fact: "In cases of male hermaphroditism, the total impossibility of surgical correction as a male is marvelously exemplified in those cases, namely of the testicular feminizing syndrome, in which nature herself produced the complete paradox of a perfect set of female external organs" (1965d, 185).

Money's penis obsession might seem to suggest a simplistic view of what constitutes a man, but Money was not concerned with manhood. He was interested in that amorphous thing called male identity. In 1963, in an article titled "Developmental Differentiation of Femininity and Masculinity Compared," he wrote,

> Hermaphroditically ambiguous external genitals may, however, tell a lie to their owner, as well as to other people, so that it is therapeutically highly desirable to have them surgically corrected at an early age. The internal organs, by contrast, being hidden, do not appear to have any effect on the differentiation of gender role and identity. (1963b, 56)

The reference to "internal organs" shows that Money saw the problem as not something that might be called "real sex" or even "real gender." He sought the psychological well-being of anyone who cannot find a non-ambiguous gender identity. The issue was not to make anyone into a man or a woman but rather to make each person believe in his or her own male or female identity. It is quite reasonable to assert today that genitals are not essential to that identity, and trans men in particular have written many essays to that effect. Whether that was the case in 1963 is a different matter.

Contrary to Colapinto and other attackers, Money's "hit and miss basis" – an interesting slip for what is usually "hit or miss" – extended to Johns Hopkins itself. His colleagues and assistants recalled how difficult it was for Money to get patients. They came to him only if referred by endocrinologists and urologists, and, even if they came, there was no way of requiring them to return for follow-up. The therapeutic interviews represented in Money's dissertation took a great deal of time. In his later work, he attempted to have these interviews with all members of the family. Money as both clinician and researcher was often thwarted in his attempts to help and to learn. Given the nature of intersex, the number of research subjects could never be large, but given the circumstances at Johns Hopkins they were fewer still. As the subjects were periodically lost to follow-up, Money's beloved longitudinal studies, which became in many ways his raison d'être, were always difficult to maintain.

These factors combined to make Money's protocol so important to him. His devotion to his research and his patients coalesced with his always large ego to make him ever the spokesman for a theory that was far from his alone. As many of his colleagues noted, his relationship with his most supportive colleagues and staff was always fractious. Whenever this protocol was in any sense questioned in print, he was always quick to pen an angry rebuttal. Colapinto notes one opposing publication by a team led by Daniel Cappon that was published in 1959. Money's papers include a letter to a colleague about this article: "I consider his effrontery in producing such a paper an academically personal one, for I am the originator, primary investigator and senior author of the work on the psychology of hermaphroditism here at Johns Hopkins. I have a sheaf of notes ready to write a rejoinder should Cappon have the further effrontery to publish his ill-thought, knavish statements." In 1970, pediatrician and psychiatrist Bernard Zuger wrote an article titled "Gender Role Determination: A Critical Review of the Evidence from Hermaphroditism." Money responded to Zuger in an article titled "Critique of Dr. Zuger's Manuscript," published in *Psychosomatic Medicine*. The tenor of his response is visible in his opening sentence: "It is difficult for the seeing to give art instruction to the blind" (1970a, 463). Later in the article, Money gave his justification for the strength of his reaction:

> What really worries me, even terrifies me, about Dr. Zuger's paper, however, is more than a matter of theory alone. In print it will represent not simply an intellectual debate between academic colleagues concerning psychosexual theory, but will be used by inexperienced and/or dogmatic physicians and surgeons as a justification to *impose* an erroneous sex reassignment on a child or adolescent, omitting a psychologic evaluation as irrelevant – to the ultimate ruination of the patient's life. I have, in the course of my 20 years of experience in hermaphroditism, witnessed the tragedy of young persons' lives ruined, in this respect. By like token, I have seen also the tragedy of refusal of a sex reassignment in selected cases of youthful hermaphroditism whose gender identity demands it. Contrary to what Dr. Zuger and other critics, including Cappon and Diamond on whom he leans for support, imply, I published in 1955 a recommendation for sex reassignment in properly selected cases, and have never backtracked on that recommendation. (1970a, 464)

In other words, sex reassignment would neither be enforced nor denied but would take place only "in properly selected cases," presumably as defined by Money. More important, probably, is that he "never backtracked." The refusal to backtrack is at once an asset and a liability. Greg Lehne recalled to me a party at his house, where Cheryl Chase, perhaps the best-known intersex activist, and Money had a heated argument about early surgery for intersex infants. Afterwards, Lehne suggested to Money that most of his theories were quite compatible with the approach Chase supported, with the one exception that Chase rejected early surgery. Lehne pointed out to Money that both society and surgical methods had changed greatly in the thirty some years since the protocol had been established. He suggested that Money should accept the changes demanded by the intersex activists and modify his protocol. Money never backtracked. As a result, he lost the support of the intersex community that he had so long championed.

While surgery might seem to be a sine qua non for transsexuals, many who identify as transsexuals have had no surgical intervention. Surgery is viewed as completely the choice of the transsexual person: surgery if necessary but not necessarily surgery. In quite a contrast, the rejection of surgery is a sine qua non for most who identify as intersex. I cannot claim to have read everything on this topic, but I have read nothing by those who could be called "intersex activists" that justifies any surgery. In the video *Is It a Boy or a Girl?* (Ward 2000), a man who had a severe hypospadias, in which the urethral opening is on the underside rather than at the end of the penis, seems to suggest that the limited success of the series of operations he had to correct it represents the evils of unnecessary intervention in intersex conditions. Part of the problem, presumably, is that surgery for hypospadias was much less effective when he was young than it is today. Regardless of his discomfort, it is difficult to see what alternative he had. Most urologists would assert that a severe hypospadias, *severe* implying either an unusually large opening or one very low in the penis, requires surgery. Still, what constitutes "severe" and what surgery is "required" are among the many controversies about intersex. I do not know the details of the man's condition, and, even with the details, medical consensus would not exist.

But, again, why was Money the focal point for so much of the antagonism against intersex surgery? Even today there are few websites devoted to intersex issues that do not contain at least one attack on Money. Yet Money was not the surgeon doing the cutting. Moreover, he was only supporting the continuation of a practice that had been well established before he was connected to any hospital. First, Money had a huge ego, as suggested by his description of himself as the "originator." He was, rather, the theorizer and the proselytizer. Second, however, he became such a presence in sexology that it was logical that this procedure be attributed to him. The title of the 1977 *Johns Hopkins News-Letter* article that I quoted in the introduction bears repeating here: "Doctor Money Speaks, the Whole World Listens" (1977e). Third, that presence included statements on issues such as child sex and pornography. Colapinto makes a number of references to Money's more infamous ideas, such as those on the sexual liberation of children, to attack him as a sexual deviant. In other words, if something is to be attacked, labelling it with Money's name is likely to increase the animosity.

Money's fame was such that he was easy to misrepresent. Colapinto ([2000] 2006, 34) claims that Money saw "human newborns as total psychosexual blankslates." Even Fausto-Sterling (2000, 3) begins *Sexing the Body* by stating, "In 1972, the sexologists John Money and Anke Ehrhardt popularized the idea that sex and gender are separate categories." In a footnote, she provides a more complex sense of the argument in *Man and Woman, Boy and Girl,* but the effect is still to suggest that Money gave short shrift to biology. In her 1979 book *The Transsexual Empire*, Janice Raymond presented the opposite view of Money. As noted in Chapter 3, *The Transsexual Empire* has been viewed by most in the transgender community as the classic example of transphobia, and thus it is not surprising that Raymond wrote extensively about Money. She admitted that "Money is no biologizer of the *ancien régime*" ([1979] 1994, 45) but asserted that the results are just the same: "Under the guise of science, he makes normative and prescriptive statements about who women and men are and who they ought to be" (44). She said that he maintained that "socialization is destiny" (45) but that socialization can happen only one way, "which takes on all the force of a new natural law" (45).

Raymond ([1979] 1994, 45) quoted Money at length, but she admitted that she needed to read Money carefully to come up with her interpretation: "While it may seem that I am equivocating with the explanation and critique of Money's theories, this is because Money himself consistently equivocates." Perhaps because Raymond was so opposed to transsexuals, she seemed unable to accept that what she perceived as equivocation was rather a careful balancing of the different forces that are part of being human. As early as 1963, Money offered a much more nuanced view:

> The simple dichotomy of innate versus acquired is conceptually outdated in analysis of the developmental differentiation of femininity and masculinity, which is not to say that one should obliterate the distinction between genetics and environment. Rather, one needs the concept of a genetic norm of reaction that defines limits within which genetics may interact with environment, and, vice versa, of an environmental norm of reaction that defines limits within which environment may interact with genetics. Then one would speak of a norm of interaction when genetics and environment are in conjunction under optimal circumstances. Abnormality, or deviation from the norm of interaction, may be engendered by alteration of either a genetic or an environmental factor to be other than optimal. (1963b, 51)

In other words, genetics are limited by what an environment makes possible, and environmental influence is limited by what genetics make possible. The individual who results is an intersection of those possibilities, not a "blank slate" but a complex intertwining of competing and complementing force fields. It is not that Money was requiring the human to fit the normative expectations that Raymond disliked but that these expectations simply were the "normative," and thus all are socialized by them, no matter how much we might wish to fight them.

But there seems little question that Money believed that the only accurate view of that complex intertwining was his. One of the most forceful arguments for the truth of biological sex for intersex individuals was an article by Julianne Imperato-McGinley and co-authors published in *Science* in 1974. Imperato-McGinley, an endocrinologist at Cornell, had done extensive research in Mexico with a group of men who were called *guevedoces,* which literally translates as "testicles at twelve"; the guevedoces were females who became males at puberty, the same condition experienced by Cal in the novel *Middlesex* (Eugenides 2003). In a response published in *Science* in 1976, Money suggested that the men had not lived normal girlhoods to then become males at twelve, but rather grew up with the community always being aware of the possibility of a change that had been experienced by other children before: "The authors disregard the post-natal, social stage of gender-identity differentiation as though it does not exist" (1976b, 872). Imperato-McGinley and colleagues (1976, 872) responded, "Money obviously does not know how the parents raised these children." Of course, ultimately, neither did Imperato-McGinley, unless she was hiding in their houses all those years. She knew only their reports of how they raised their children. However, the argument reveals much about Money, who could so easily dismiss Imperato-McGinley's work because he knew what would have happened in a situation of which he had no specific knowledge. Yet this also represents Money's absolute belief in the error of an assumption of the imprimatur of biological sex. In a 1973 review of Corinne Hutt's *Males and Females* (1972), Money rejected her claims for biological determinism and asserted that, "if taken seriously, this book will doom human sex research in Britain to remain exactly where it is, namely, hypotrophic. Shame on Penguin Books for publishing a paperback so misleading in its scholarship" (1973a, 604).

In his 1986 book, *Venuses Penuses*, Money mentioned his search for "conceptual order" (1986d, 11). Arguably, Money's desire for conceptual order often overcame the necessity for pragmatism, as in the argument with Cheryl Chase. The limitations of his field, the small number of intersex individuals available for study, the limitations of his position at Johns Hopkins, and the difficulty of follow-up made it impossible to have the range of patients necessary to prove any theory. Add to this his sense of a cause – to protect intersex individuals from "inexperienced and/or dogmatic physicians and surgeons." The result was his all too immediate reaction to see any opposition as simply inertia or refusal to respond to the need for change.

Early-twenty-first-century commentary on intersex seems to suggest that Money is part of a benighted past that has been superseded. A 2006 meeting of a group of twenty-nine international human rights experts, in Yogyakarta, Indonesia, created the *Yogyakarta Principles on the Application of International Human Rights Law in Relation to Sexual Orientation and Gender Identity,* which they released in a press statement and online on March 26, 2007. According to the group's website, the Yogyakarata principles include the following:

> Take all necessary legislative, administrative and other measures to ensure that no child's body is irreversibly altered by medical procedures in an attempt to impose a gender identity without the full, free and informed consent of the child in accordance with the age and maturity of the child and guided by the principle that in all actions concerning children, the best interests of the child shall be a primary consideration.

Of course, these principles have no legal effect in any individual jurisdiction. They suggest a procedure taken in response to a situation such as the Reimer case rather than to an extreme intersex condition, in which either gender is in some sense imposed on the child. Any operation is invariably made with what those involved consider "the best interest of the child." To suggest that no intersex condition should ever be treated surgically goes into the realm of questions about when surgery is necessary and what conditions represent a disability. Just as one example, cochlear implants for deaf children might be seen as necessary surgery to correct a disability, but many in the deaf community disagree. Some intersex activists have suggested that surgery should only be in cases that are "life-threatening." To many, this position would seem extreme: many other conditions that are not life-threatening are corrected as a matter of course, such as a cleft palate. And, of course, even the term *life-threatening* is open to broad interpretation.

The problem with Money's protocol was not his own view of what could be normal. His dissertation showed that "normal" could be someone whom medicine deemed to be quite abnormal. But Money believed that "normal" would be difficult to achieve for someone whom society could not deem to be normal. Society has not changed as much as we – and he – might have hoped. It is still difficult to see answers that do not require extensive social change. Today the term *intersex* is in reasonably common usage and does not simply represent "freak" as did *hermaphrodite* in an earlier era. However, the understanding has not necessarily shifted significantly. Just as one rather arbitrary example, in 2004 the *Calgary Herald* published an article, "Neither Boy Nor Girl," on Ilizane and Xenia Broks, two half-sisters with CAIS. It is primarily an interview with the two girls and their fathers, but the description is telling: "Outwardly they look like girls: in truth they are half and half" (Craig 2004). Apparently, this is how they refer to themselves as well, but it is, of course, completely inaccurate. If one relies on chromosomes, they are simply male. If one relies on appearance, they are simply female. They cannot reproduce, like many other females. They have moments when they consider their behaviour rather masculine, like every other female. To call them "neither boy nor girl" seems to suggest a category not really human. To call them "half and half" is to turn a medical condition into a science fiction movie. Or perhaps it is the equivalent of *deaf and dumb* or *cripple* or any other term that takes a medical fact and turns it into some weird thing that separates a person from every other human. Activists for all medical conditions have made it their first task to establish the humanity of all individuals. On the Intersex Society of North America website (2008), Cheryl Chase cautions us to "Label a medical condition, not a person." Her advice is particularly applicable here. Some might wish to except from Chase's directive what has been called the "true hermaphrodite" – a person with significant genitalia and reproductive organs of both sexes – someone who is perhaps truly half and half, who is not just a male or female with some differences but a womanman or manwoman. Perhaps this is the person who goes beyond the condition. The true hermaphrodite has been the intersex figure most loved by popular culture. The title character in Kathleen Winter's 2010 novel *Annabel*, for example, is a true hermaphrodite, but, like most fiction writers who represent true hermaphrodites, Winter's main interest is in presenting a person whose psychological attributes are evenly split between male and female, as are his physiological attributes. Winter suggests that he has even impregnated himself, a constant narrative in the popular imagination but something that has never been recorded medically. There is no record of any person both impregnating someone and being impregnated him/herself, which presumably would be the default category of a true hermaphrodite. Case histories and autobiographical memoirs suggest that true hermaphrodites are not even psychologically bi-gendered. Studies such as Young's *Genital Abnormalities* (1937) show that true hermaphrodites are people who live in one gender and find themselves having some attributes of the other. They are like all of us.

But many intersex individuals choose to be not like all of us. Certain communities choose to live separately from those who are different, from religious sects to lesbian separatists. This should be their right. If intersex individuals wish to identify themselves as intersex, so be it. Intersex activists have worked strenuously to protect intersex infants from unnecessary intervention and intersex adults from discrimination. Like all persons with complex medical conditions, they need particular consideration. Ultimately, however, like diabetics or cancer patients, most simply wish to be normal individuals, not identified as some medical category.

This is not a book about these conditions but about John Money's participation in them. Thus, the book is primarily historical. The history of medicine shows a significant shift from the Second World War to the beginning of the twenty-first century, from a society in which doctors tended to tell patients what must be done to them to a society in which patients wish to treat doctors as resources to enable them to live life as they choose to live it, regardless of medical opinion. The most coherent statement about today's ways of fixing sex is from the Intersex Society of North America's (ISNA) website (2008):

> In 2007, ISNA sponsored and convened a national group of health care and advocacy professionals to establish a nonprofit organization charged with making sure the new ideas about appropriate care are known and implemented across the country.
>
> This organization, Accord Alliance <http://www.accordalliance.org>, opened its doors in March 2008, and will continue to lead national efforts to improve DSD-related [disorders of sex development] health care and outcomes. Accord Alliance believes that improving the way health care is made available and delivered is essential to ensure that people receive the services and support they need to lead healthy, happy lives.
>
> With Accord Alliance in place, ISNA can close its doors with the comfort and knowledge that its work will continue to have an impact.

In other words, rather than intersex activism, the chosen approach is a community of professionals who advocate the *Yogyakarta Principles* noted above.

When Money retired, the psychohormonal research unit was disbanded. Needless to say, this devastated Money. He concluded *A First Person History of Pediatric Psychoendocrinology* with this sentence: "There is always the unpredictable chance that, phoenix like, a psychohormonal research unit will rise again from the ashes of its own fire" (2002a, 126). Regardless of what was and was not done, his intentions were clear. In a 1990 interview for the *Baltimore Sun*, he remarked,

> There is such an unbelievable taboo on sex … I always tell parents who ask me that as far as I can make out from what I know, without any actual statistics ever having been gathered, a birth defect of the sex organs is about as common as a birth defect of the mouth, as in a harelip. But we certainly aren't able to deal with a birth defect of the sex organs with the same degree of equanimity as we deal with a harelip. (1990c)

It is worthy of note that *harelip* is no longer used. The term is regarded as offensive because it suggests that the person with such a lip looks like an animal. The current term is *cleft lip*. Even in 1990, *harelip* would have been unlikely to appear in anything for a medical audience. So, was Money just out of date, or was he once again using language that represented how the world understood something rather than how his peers might? In any case, Money's failure of language should not hide the accuracy of his observation. It is sad that "equanimity" with other medical problems was clearly the purpose of all of John Money's clinical work, research, and publications but that today he is remembered rather as an imperious medical interventionist. To fail to recognize the generosity of his intention is to fail to recognize so much.

CHAPTER THREE

LOVEMAPS

The Relationship World

Money was always reworking language – and not always to his benefit. The word *lovemap* is a good example of Money's word-play. The sound of the term seems to resonate with self-help doctors on after-noon television. In fact, the ubiquitous psychologist John Gottman (1999) used something he called the "love map" in one of his marriage manuals. However, Greg Lehne, a colleague of Money whom many regard as the most knowledgeable follower of his theories, sees the lovemap as one of Money's most important contributions. Money's own belief in its significance can be seen both by the number of times he reused the concept and by the fact that he published three books with this word in the title: *Lovemaps* (1986c), *Vandalized Lovemaps* (1988b), and *The Lovemap Guidebook* (1999a). The guidebook was meant "to appeal to the reader with a personal curiosity about lovemaps in everyday life and relationships" (1999a, 8). Perhaps Money had hopes of competing with Gottman and the other relationship therapists.

Mapping is a way of understanding things spatially. The most familiar form is the two-dimensional representation of geography, but there are many kinds of maps. The term is often used for diagrams of the body, particularly the brain, as well as other aspects of science. Mapping represents relationships in ways that contain and exclude. The map provides a whole in which all ele-ments are somehow in connection with each other. Yet the map also neglects extraneous elements, and the undeclared absences can be equally revealing. For instance, although some originally thought that the mapping of the human genome might create a complete representation of what a human is, most believe that the results are decidedly insufficient for such a claim. Much of what a human is lies outside the map. Terra incognita continues.

As a term, *lovemap* is silly, but it is also insightful. In the introduction to *Lovemaps,* Money referred to the term as a coinage to serve a purpose in lectures at Johns Hopkins:

> In many instances, a person does not fall in love with a partner, per se, but with a partner as a Rorschach love-blot. That is to say, the person projects onto the partner an idealized and highly idiosyncratic image that diverges from the image of that partner as perceived by other people. (1986c, xv)

Money went on to explain,

> To communicate fluently with students, I found it extremely awkward to have only the expression, *an idealized and highly idiosyncratic image.* Therefore I began substituting the single term, *lovemap.* (1986c, xvi)

In other words, each person's lovemap provides the sexual aims and sexual objects that will be desirable for that person. He prophesied, "Sooner or later, therefore, *lovemap* will find its way into the standard dictionaries of the English language, and in translations" (1986c, xvi).

Money's hubris was always visible, and, while the last claim was clearly wrong, he might have been quite correct about the value of the concept. At the normative level, the concept of the lovemap might seem to be making a point oft made before, as in the various aphorisms that claim that a lover sees the beloved through different eyes. If one looks beyond the normative, however, the lovemap becomes an explanation for any sexual deviation, as an object not usually perceived as sexual becomes the object of desire. By using the word *love* to include all aspects of desire, Money was able to consider the shoe fetishist within the same concept as the Cinderella who finds her prince charming. Both are responding to their lovemaps. The loveblot, the object to which the lovemap responds, can be a person actively pursuing his or her own lovemap or an inanimate object, or it can even be an act with no obvious object of any sort, such as for those whose lovemap finds erotic arousal in

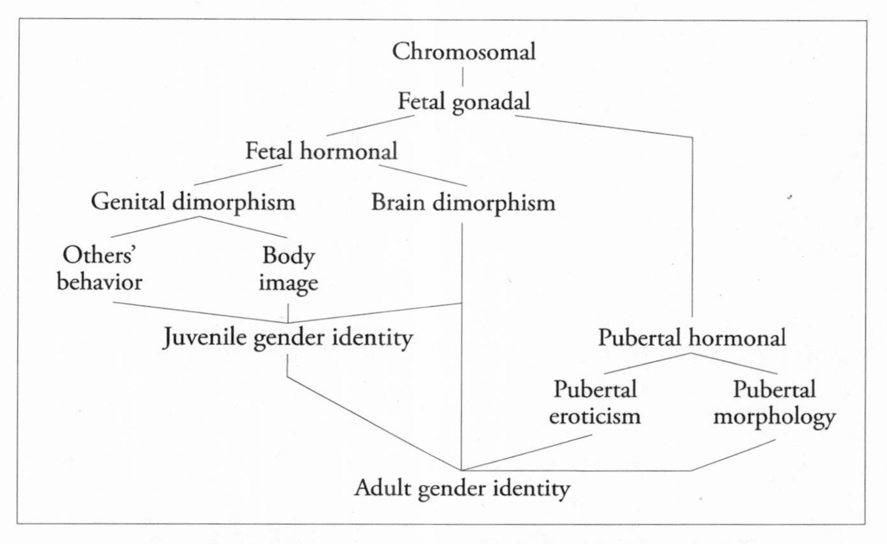

Diagram illustrating the sequential and interactional components of gender-identity differentiation (Money 1972d, 3)

breaking and entering buildings but without any further interaction with the inhabitants or their possessions.

While he later referred to it instead as a *gendermap,* Money's "diagram to illustrate the sequential and interactional components of gender-identity differentiation" in *Man and Woman, Boy and Girl* (1972d, 3) represented the process through which the lovemap is created. Lehne suggests that the great number of reprints shows that the diagram represents Money's greatest influence on psychological theory. It does not provide any of the details of the lovemap as set out in his later books, but the diagram shows how the stages interact as the lovemap develops. It also suggests the various crucial points at which the lovemap is formed, solidified, and, in some cases, skewed in response to the different aspects of human maturation (see diagram).

The process of the diagram is chronological, from the top, but it attempts to overcome the typical view that human development begins with nature and then after birth becomes nurture. Instead, the moment of conception – and any time before – is "chromosomal." Then the next stage, in the womb, is "fetal gonadal." The hormonal development moves rapidly to "fetal hormonal." It takes much longer to reach "pubertal hormonal," but in any person not given additional hormonal treatment the movement is the same. After hormonal development in the fetus comes genital dimorphism and brain dimorphism. There is much controversy over brain sex, with some theorists seeing a strong biological difference between male and female brains. A good response to such claims is offered by Cordelia Fine in *Delusions of Gender* (2010). "Brain" in this diagram refers both to the physiological brain and to what might be called "mind," the realm of thought however it works neurologically. In this diagram, both genital dimorphism and brain dimorphism happen at the same time, but they are quite separate, so one should not assume that genital dimorphism causes brain dimorphism or vice versa. According to the diagram, "body image" and observation ("others' behavior") extend not from brain dimorphism but from genital dimorphism. Thus, central elements are the person's recognition of his or her own genitals and the reaction by others to those genitals or to the person's own understanding of his or her own genitals. This suggests why Money saw genital conformity as so important for intersex cases. Only at the stage of "juvenile gender identity" does brain dimorphism return to be an influence. This is potentially very interesting for the transsexual as it seems to suggest that a child can develop clearly as a male while the brain is developing as a female. There will inevitably be a conflict as the brain to this point has not had control over "body image" or "others' behavior." At the point of puberty, all the factors come together to create the "adult gender identity." It must be noted, however, that while "brain dimorphism" is linked to "juvenile gender identity" it also has a direct line to "adult gender identity." While most transsexuals claim to have known at a very early age that they were in the wrong gender, Money's model provides a clear possibility that certain individuals will develop along the expected trajectory until adulthood, when the difference in brain dimorphism becomes clear.

From the lovemap, Money moved on to the gendermap. As he explained in his 1995 book, *Gendermaps*, the term is a "cognate of lovemaps" (10), but it seems more of an overlapping of two fields, perhaps a Venn diagram. Clearly, gender identity forms long before anything that might be called a lovemap is visible; yet, for every adult, the developing intersections between the two can be seen from a very early stage. Money stated,

> There's no bypassing the gender identity fork. It is practically impossible for a person to develop any sense of identity at all without identifying as either a male or a female, and the gender identity gate locks firmly behind. Yet in a very special way, this fork must be straddled. You had to construct both male and female internal models – schemas – in your brain, concepts of what it means to be male and what it means to be female. (1975c, 88)

This "gender identity gate," which occurs at about eighteen months old, became central to Money's understanding of gender and sexual identity. The connection between gender identity and lovemap might seem to be only the division between heterosexual and homosexual: your lovemap could take any shape, but most would believe that a central aspect is the process that decides whether it will be oriented towards an object of the same sex or an object of the opposite sex. Money instead saw the two continually enmeshed for all humans. The conscious and unconscious sense of the self as male or female constantly shapes erotic possibilities. And more. In one of his last books, *Unspeakable Monsters in All Our Lives,* Money (1999b, 21) asserted the inevitable omnipresence of the dimorphic choice: "The maps and programs of mindbrain functioning are matched in polarized pairs, one negative, one positive. Thus, for example, every individual possesses a gendermap for masculine and one for feminine, one of which is coded 'mine,' and the other 'thine.'" Some might be irritated by the apparent heterosexism of that "thine." This is not "yours," or that which is simply "other," but "thine," which seems to imply a particularly intimate other. Money's probable answer to this observation would be that heterosexuality is so entwined in our cultural understanding that it would be impossible for anyone, no matter how homosexual, to understand the opposite gender without an awareness of that prospective intimacy.

This understanding justifies his belief in the omnipresence and centrality of gender dimorphism, but it still might seem contradictory given that he had very limited belief in anything like scientific truth in that split. There were many glimmerings of feminism in Money, not least because he saw reproduction as the only aspect of sex difference that is essential in the largest sense, as he noted in *Gendermaps* when commenting on the various studies that assert larger barriers between male and female:

> In the trappings of modern science, these depictions together reincarnate the earlier pejorative stereotype that depreciates the female's intellect and aggrandizes the male's as superior, in support of the erstwhile shibboleth that it is foolish for the female to compete academically and vocationally with the male. Far from being phyletically determined, this shibboleth of male/female difference is the product of cultural and political overlay. (1995, 48)

From the time of his first intersex studies, Money was convinced that the sex drives of male and female are not different. This is one of many of his theories that seemed rather avant-garde when he first postulated it and then was maintained without significant change throughout his career. In 1960, he wrote a chapter in a collection titled *Recent Advances in Biological Psychiatry.* He summarized his claims:

> The level of sex drive or libido is hormonally influenced, and androgen is probably the libido hormone in both men and women. The direction or content of the erotic inclination in the human species is not controlled by the sex hormones. Hormonally speaking, the sex drive is neither male nor female but undifferentiated – an urge for the warmth and sensation of close body contact and genital proximity. (1960a, 223)

Thus, the lovemap is a psychological shaping of this sex drive. The lovemap can incorporate a variety of strange deviations. While they are propelled by the sex drive, however, it is not the sex drive that defines where these deviations will go. One repercussion of this theory was the development of Depo-Provera, an anti-androgen, to treat sex offenders. It was first used under the supervision of Money and his colleague Claude Migeon in 1965. Depo-Provera is often called "chemical castration" not least because of its association with birth control and its tendency to produce feminizing side effects. Money refused the term as a complete misnomer. As he pointed out, the purpose was not to demasculinize or correct excessive masculinity but to decrease sex drive and thus limit criminal behaviour; Depo-Provera was "used for the treatment of sex offenders with paraphilias of the illegal type" (1987b). Money was well aware that the vast majority of sex offenders were male, but he did not see the chemical restraint of their urges as about gender.

One aspect of the lovemap that is of particular interest is the refusal to separate love and lust. Money saw the split as a product of history. In *The Adam Principle,* he explained,

> In the Christian era, love was assigned to the spirit, and lust to the flesh. From the eleventh century onward, under the influence of the troubadours, love was secularized as romantic affection. It was destined to be forever unrequited and unconsummated in genital lust, for the fair lady's hand was always either promised or already given in an arranged marriage. The split between love and lust has haunted occidental civilization ever since. It permeates everything

> ... In sexology as science and medicine, sex has replaced lust, terminologically, and attraction has replaced love. (1993a, 17)

Yet, for Money, one cannot come without the other. In one of his many interviews in what might be called the soft-porn media, in this case in *Gallery* magazine, he said,

> I don't think there's ever sex without romance. It may be very truncated, but there's always some kind of emotional component in what happens between two people sexually. I guess the most extreme example of "impersonal" sex – that's a silly term because sex is always personal – the most extreme example of casual sex is when people meet on either side of a dividing wall in a toilet through a glory hole, but even that's pretty emotionally fraught with excitement and intense feeling for the people who get involved in it, even though it's totally anonymous. (1977k, 57)

Although he never commented on this aspect of his personal life, Money had significant experience of casual sex, with both men and women. Many would find the suggestion of romance in stranger sex ludicrous or even offensive. However, Money was trying to see something closer to a complete picture of sex and emotion. Any romance aspires to love:

> Love means pair-bondedness. It may be sacred or profane. Profane love may be erotic or affectional. Erotic love may be recreational or prorecreational, connubial or companionate. Affectional love may be filiative, neighborly, or comradely. Pair-bonds vary in strength and duration. The longest-lasting pair-bond is the mother/child and/or the father/child bond. The most intense pair bond is that between two lovers. Its highest passion typically has a maximum duration of two or three years and then wanes. Love may occur at first sight, or later. It resembles imprinting. The love may project onto the partner an idealized but idiosyncratic image, as is routinely done in sacred love wherein the partner is metaphorical. If the erotic partner fails to live up to the projected image, the pair bond pulls asunder, and love may turn to hate. Reciprocated bonding, by contrast, is durable, and easily expands to encompass parent-child bonding. (1978e, 14)

Money's comment about "sacred love" suggests how widely he was willing to expand his category of "pair-bondedness." *Pair-bonding* is normally used by biologists who study animals and is seldom employed for humans. To apply it to humans might seem to reduce them to animals, but then to apply it to sacred love might seem just confusing. A love for God is a pair-bond? Money's purpose, as with the lovemap, was to understand the dimensions of human romantic and sexual attraction scientifically. In an interview in 1976, he called for more openness in sexual science and suggested as one possibility, "If you want a nice, easy way to talk about love and get it into science instead of poetry, call it pair-bonding. Then you can study it in animals, as well as in people" (1976c, 11). For many of us, such mapping might seem reductive, but to Money it was part of an attempt to expand known knowledge to be ready to encompass unknown knowledge, to take the map of the known world and make a map of spaces yet to be drawn and perhaps even yet to be seen. In Money's view, to see romantic sex between cohabiting partners as something wholly different from sex with a stranger misunderstands the complexity of both. As so often in Money, the normal and the abnormal were less separated in kind than by placement on the map.

The term *imprinting* perhaps takes matters even further. Money used the word from very early in his career: it appeared in one of his first publications with the Hampsons in 1957, "Imprinting and the Establishment of Gender Role," in which they compared gender role to Nobel Prize-winning ethologist Konrad Lorenz's classic account of the imprinting of grey geese. Money used a similar argument in a paper that was more specifically about a precursor of the lovemap:

> Falling in love resembles imprinting, in that a releaser mechanism from within must encounter a stimulus from without before the event happens. Then that event has remarkable longevity, sometimes for a lifetime. The kind of stimulus that, whether it be acceptable or pathological, will be the effective one for a given individual will have been written into his psychosexual program, so to speak, in the years prior to puberty and dating back to infancy. (1972a, 213; 1986c)

While many have criticized applying imprinting to human behaviour, the specific compulsion that many humans feel towards a specific type of sexual object, a compulsion that can be obsessively pursued again and again with many objects that fit the type, seems very close to the classic understanding of animal imprinting.

One example of Money's attempts to comprehend the dimensions of human sexual attraction was his constant use of the word *limerence*. For once, the neologism was not his own, but he seemed to find the same fascination with it as with his own words. He offered many similar definitions of the term. This one appears in *Gendermaps:*

> *Limerence* is the term that signifies the state of being love stricken. It may also signify the state of being love sick. The term *limerence* was coined by Dorothy Tennov in 1979. Being limerent appears to be a peculiarly human experience for which there is no easily accessible animal model. (1995, 113)

Even among animal species that mate for life, biologists would hesitate to use a phrase such as "love stricken." All human cultures, however, have believed in being love stricken, and much of art depends on it. From Tennov, Money found a word that could encapsulate this feeling that is more physiological than "love" and yet is much more complex than the kinds of sexual response identified by a plethysmograph, which measures changes in blood flow in the penis to assess sexual arousal, and similar strictly physiological assessments.

In a study of paraplegics who had no genital response, Money found that their dream orgasms were much the same as for those who had normal functions: "It offers conclusive evidence that cognitional eroticism can be a variable of sex entirely independent of genitopelvic sensation and action" (1960c, 21). This seems to prove what many have said and what Money always believed: the brain is always the most significant sex organ. Limerence also goes beyond the simple couple-based ideal so often meant by *love*. In one essay, Money referred to sadomasochism as "pathological limerence" (1997, 542). In other words, the category for the desire felt by the sadistic or masochistic subject is the same as it would be for a pair-bond desire that our culture would deem acceptable. Money could stretch this category to quite an extreme degree, as when he referred to the Stockholm syndrome, in which the kidnapped develops a relationship with the kidnapper, as a "limerent love affair" (1986c, 91).

Money had a similar purpose in his devotion to using the term *paraphilia* to replace what Freud and others called *perversions*. As Freud noted, perversions are often in the eye of the beholder and are most accurately defined as simply practices that society finds unacceptable. Although the *Oxford English Dictionary* records the first usage of *paraphilia* in 1925, it was largely Money who popularized the term among psychologists. Eventually, the word replaced *perversions* in psychiatric literature. Money's primary interest was to confirm that paraphilia is identified solely by its relation to society, as noted in a lecture he gave to Johns Hopkins medical students in 1970: "The specificity of the sexual arousal pattern of most people who qualify either as paraphiliac or as merely erotically eccentric, resembles in its delimitation that of the so-called normal person" (1970d, 25). In other words, the object and action, no matter how strange they might appear, usually follow the same trajectory as the types of sexual desire that are most universally accepted. He defined paraphilia as any sexual desire that is not accepted as *normaphilia*, or sexual desires that society finds normal. Perhaps not surprisingly, *normaphilia* does not yet appear in the *OED*.

Money's analysis depended on an absolutely environmental judgment that the paraphilia is identifiable only as that which society deems to be not normal. At the same time, however, he looked constantly for biological reasons for deviations in sexual desire. One of the more extreme versions of this search appeared in a comment in the same Johns Hopkins lecture on persons who exhibit a number of paraphilias. The language might seem a bit confusing. He was referring to a very small category of person, someone whose lovemap is not tied to one or two paraphilias, someone who has an ardent and aggressive desire that manifests in a number of apparently arbitrary directions:

> Developmental nonspecificity of sexual activity is of unknown etiology, for the most part. It differs from the unexpected and sometimes sudden loss of specificity that may accompany a brain lesion of adult onset, or the cerebral deterioration of certain types of senility. There is some incompletely confirmed evidence that developmentally impaired sexual specificity may be a function of an abnormality in the genetic code, particularly in the XXY and the XYY syndromes in phenotypic males (see Introduction). For the most part, however, the detailed etiology of the condition whereby a plurality of phyletic mechanisms become enlisted in the service of sexuality in one and the same person remains to be ascertained. When this problem is scientifically solved, then the etiology in full of all the sexual behavior disorders will have been solved. (1970d, 25)

Everything in that paragraph seems completely reasonable until the end. When I first encountered this passage, I found myself rereading that last sentence again and again. I still cannot believe Money wrote that "the etiology in full of all the sexual behavior disorders will have been solved." Would that ever be possible, whatever was "ascertained"? Perhaps it was said with a sarcastic tone in his voice. It is but one line in an unpublished document, and there are many instances in which Money seemed to suggest otherwise. However, as an example of scientific hubris, it could not easily be equalled.

In the same lecture, he offered an evolutionary interpretation of one paraphilia, the masochistic desire to be urinated upon:

> The phyletic mechanism to explain this particular association is probably not associated with infant care but with a modification of the primate territory-marking trait, seen for example, in the squirrel monkey, namely displaying the penis and challenging an intruding male by urinating at or on him. (1970d, 7)

This does not make Money into a simplistic evolutionary psychologist. Instead, he was looking at what is normal behaviour in the primate to understand what is deemed to be abnormal behaviour in the human. In another lecture, he stated,

> The pathological status of anomalous psychosexual behavior derives from the fact that when a component or sequence of behavior, displaced from its original phyletic context, becomes enlisted in the service of erotic arousal, it does so not simply as an option but as an imperative without which the person's erotic arousal fails. (1970e, 1)

The term *phyletic*, again one of Money's favourites, represents the evolutionary connection to primates, but it also suggests both a historical and a taxonomic relationship. Thus, something that could have been a valid and even normaphilic category in the deep past can be displaced in the process of history, or at least society can perceive it to be displaced. The taxonomic pattern is what maintains the connection to the earlier manifestation. Money's process followed the relationship between taxonomy as a form of understanding and mapping as a form of understanding. The taxonomy establishes categories, and the map demonstrates spatial connections.

Lehne suggests that the socio-psychological problem of the paraphilia is primarily that "imperative" to which Money referred. Lehne notes that often a paraphilia that manifests as a major and even criminal problem is something that functions quite comfortably as a small part of someone else's lovemap:

> It is tempting to use a hydraulic model to explain this high energy force of arousal or motivation. If a typical lovemap includes a larger territory to explore and act upon in sexual situations, like water flowing over a broad plateau, the energy is spread out and there are many options for run off. However, if the territory is very limited, like water flowing through a small valley, the force of the sexual energy is much greater. (2009, 16)

Thus, Money turned to the drug Depo-Provera to limit the flow of water through a very small valley. The need for some such treatment can be exemplified by the case, such as Lehne mentions, of a man who has been arrested eleven times for masturbating in a public park. Almost all men masturbate, and many have probably done it at least once or twice outdoors. Even a slight and contained exhibitionism is probably not that uncommon. But very few have felt so drawn to that small valley of public masturbation to have been arrested eleven times.

Money viewed paraphilia as a "vandalized lovemap" (1986c). In other words, something had wantonly taken a blunt object and broken a perfectly good lovemap. This weapon could have been genetic, chromosomal, or hormonal, but it also could have been environmental, from the chemical to the psychological. Money consistently saw the problem, therefore, as in some sense appropriate to medical treatment, following the disease model. While the decision as to what is paraphilic is society's – that which the culture considers to be not normal – the problem of paraphilia for the person exhibiting the paraphilia is a disease – something that interferes with function. From the beginning, Money said that homosexuals could be either normaphilic or paraphilic. There is no reason to assume that a homosexual would find it difficult to function because he or she has an adult human as a sexual object, with no other socially deviant factors.

The concern for function begins with the person who has the paraphilia, but the person's dysfunctionality can then interfere with the functions of the rest of society. In the case of someone who commits rape and murder, there can be no question that the paraphilia is larger than the paraphiliac. In the case of someone who does something alone in his own bedroom, one might question whether society should have any involvement at all. In the case of the man incessantly masturbating in the public park, there is no pure and simple answer. This is one of the reasons Money quite consistently opposed criminalization of the paraphiliac, whether in his publications or in his various appearances as an expert witness. He displayed the vehemence of his feelings in one of his comments on arresting young paraphiliacs: "Treating the long-term effects of violent and sexual traumatization in infancy by the traumatization of long-term imprisonment in adolescence is ethically the equivalent of amputating the arms of an amputee as a treatment for having lost his legs in a car wreck" (1997, 549). He went so far as to suggest that the ethics of medicine might require civil disobedience:

> The dilemma is an all too familiar political one, namely, whether it is more expeditious to work within a system, endorsing its existence while attempting to reform it, or to assault the system from without, as a member of a guerrilla underground. Translated into the realities of clinical practice, the dilemma is whether to treat people whose sexual behavior is illegally sex offending, in order to help them stay out of prison, or to refuse treatment and fight for the reform of the system. (1983g, 250)

In this, as in so many aspects of sexology, Money feared that the forces of repression could harm both research and practice. He foresaw how this dilemma would play out at an institutional level; in an interview in 1990, he said, "The Sexual Disorders Clinic at Johns Hopkins Hospital may offer the nation's best treatment for pedophiles, rapists, and other sex criminals. But the clinic's tough job is not winning any friends, and the bad press may well close it down" (1990d, 42).

In his book *The Destroying Angel* (1985b), Money described, in a rather freewheeling manner, what he saw as the pernicious effect of the Christian moralists of the late nineteenth and early twentieth centuries. He believed that part of this effect was to increase revulsion at what are now called paraphilias and thus to increase the desire to criminalize them. Anything that is criminalized is less likely to be treated as a medical problem. To Money, Christianity represented antiquated claptrap with a wide range of errors: "There is no word in the Bible for falling in love or being love-smitten. This absence of a word for the very dramatic human experience of becoming erotically pair-bonded perhaps reflects the fact that, in the ancient times of the Old Testament, the marriage of a daughter or a son was arranged by the families" (1985b, 37). If the whole philosophy is based on suppressing lust and failing to recognize limerence, it can be of very little value.

Sexual degeneracy theory may not depend on Christianity, and sexual repression may not depend on sexual degeneracy theory, but Money saw them as closely related. He believed that such ideologies had a direct effect on the possibility of normal development in early childhood: "These are the years when sexual degeneracy theory, outmoded but still socially influential, lurks in a child's life like a polluted smog in biological warfare and sabotages the lovemap instructions that the brain receives. Consequently the lovemap gets misprinted, and the misprinting may turn out to be a paraphilia" (1985b, 138). The term *misprint* fits the idea of the map and suits Money's belief in imprinting. For Money, everything always connected.

Money's attraction to analogy was useful in that it made complex new ideas immediately understandable, but it was also reductive, as is any analogy. His constant reference to homosexuality as like left-handedness is a perfect example. It is difficult to see ways in which left-handedness is like homosexuality except that historically society did not accept either. Even just to admit that both are innate is probably overstepping the mark. The analogy fails before it starts. While he never, to my knowledge, used the term *universal grammar,* that concept seems to be behind his frequent analogy of the lovemap to a native language. Just as Noam Chomsky maintains that each human is born with a "language acquisition device" (1965), a space waiting for language, and it is just a matter of time before a native language is poured into that space, so too Money saw the lovemap as a human given: "After birth, as in the case of native language, an extraordinarily large amount of behavioral sexual dimorphism still remains to be programmed into the developing system" (1978c). Of course, Money being Money, he took this idea much further, as he demonstrated during an online chat in 1996:

> The new idea is about evolutionary sexology, which itself is a new development. Briefly stated, the idea is that the songmap is the evolutionary bridge that contacts the lovemap and the speechmap. From that it may be conjectured that the very first human speech was a lovesong.

Thus, evolutionary psychology and universal grammar were linked to Money's lovemap and moved then to song. He then used this link both to connect humans to animals and to differentiate them from animals:

> Humans before they develop true speech had like their animal cousins been able to communicate with chattering, hoots, whistles and so on. Grunts, screams and yells, however, do not lead to true human language with verbal dialogue, pure reason and numerical calculation. The origins of human speech must lie elsewhere. Where else to

> look? My answer to that question was in singing. In other words, music and sound for which I like the word songmap. (1996)

Money believed the lovemap to be a clearly proven fact, so he was ready to move on: "So there you have the hypothesis of a connection between sex and language – evolutionary connection that waits to be confirmed" (1996).

The native language analogy fit beautifully with Money's belief in what he called "sexual rehearsal play," in which children engage in behaviour that prefigures sexual activities of adults. Money often repeated what he said in 1965: "There is a small amount of anthropological evidence (Ford and Beach, 1951) to the effect that sexual aberration is rare or absent in societies without a sex taboo, where sexual and copulatory games of children are tolerable to the adults, and where adult eroticism is not rigidly hidden from children" (1965g, 13). Money used his slight knowledge of Australian Aboriginal culture and the Batak people of Indonesia to reinforce this assertion, although, as Colapinto notes in *As Nature Made Him,* ([2000] 2006) Money was probably wrong. In any case, Money most definitely did not have sufficient information to make this claim. Regardless of the anthropological justification, however, there is a logic that arises from the analogy: children will learn a language regardless of education, but there are many instances in which education can warp this development. Around the world, schools have banned an indigenous language in the process of imposing an invader's language. As many studies have pointed out, the result is not only the loss of the indigenous language but also a variety of psychological problems (Hallett, Chandler, and Lalonde 2007). In interviews, he was willing to make quite extravagant claims about the deleterious effects of avoiding sexual play: "Once you understand these syndromes, you find out they are not what they appear to be," Money said. "Pedophiles don't come down on flying saucers from Mars. They are our children and we bring them up to be pedophiles by punishing them when we see them engaging in sexual play that's normal for children their age. Pornography has nothing to do with it" (1990e, 47). Money's argument that pedophilia is a result of punishing sexual precocity is one I have not seen before. The logic is that a child is likely to focus his sexual attentions on another child. If these attentions are punished, then the child's sexual development can be stunted, and the sexual object can be fixed at a child's level.

The theory of the lovemap enabled Money to understand a complete range of behaviour, but its base came long before the term, in what he called "exigency theory" or the "basic inevitables of being human," which he believed to be "phylisms," or an inherent part of the development of the human. He first speculated about exigency theory in "Basic Concepts in the Study of Personality" (1947), a paper given in New Zealand before he moved to the United States. It became a foundation of his understanding of psychology, as demonstrated in an essay he wrote at Harvard in 1949. His professor's response in the margin was telling: "Because of some basic trust I am giving *you* the benefit of the doubt at the point where you move from *Operations of Interpersonal Feedback* to *the Inevitables*. Please call on me soon to sustain my trust that this is *not* nonsense" (1949a, 7). Ever the big idea and ever the hesitant response from the establishment.

Money (1986d, 29) saw exigency theory as a "fundamental departure from each of the two competing philosophical premises on which most of contemporary psychologic theory is built, namely teleology construed as motivation; and mechanistic determinism construed as behaviorism or behavior modification." The former suggests that the individual is primarily driven by a goal. The latter claims that the experience of development compels the individual to act in certain ways. Instead, his new theory "brings system to the science of psychology." His list of the inevitables, in its final version, contains "pairbondage," "troopbondage," "abidance" ("continuing to remain, be sustained, or survive in the same condition or circumstances of living or dwelling" [1984a; 1986d, 115]), "ycleptance" (naming and being named), and "foredoomance" (mortality). His hope for this system was thus not causality but rather the mapping, bringing order to the categories that inevitably shape human life.

A good example of how these categories support the lovemap is found in the various ways Money moved from his understanding of intersex to an understanding of sex and gender and then to areas of study that might on the surface seem quite far removed. One example is his study of autism, based on the "phylism of troopbonding":

> In the syndrome of autism, the phylism of troopbonding is specifically impaired, and other phylisms may be hypertrophic. The phylism of limerent (lover-lover) pairbonding is not impaired, and it may be the source of discordance with other members of the troop. (1983e, 249)

In other words, the autistic individual is inhibited from connecting with the community, with the various persons from the general populace who form our immediate society, but has the usual human desire for, and ability to connect with, a specific romantic partner. From this perhaps obvious observation about the character of autism, Money moved to a very broad assertion about the way his phylisms work:

> The determinants of phylisms are neither entirely nature nor entirely nurture, but the product of the interaction of the two in varying proportion at a crucial period or periods of development. Phylismic theory is not teleologically causal and motivational, on the one hand, nor mechanistically causal, on the other. It is, however, dynamically causal, for it allows for multiple determinants to influence one another in dynamic interaction, including cybernetic or feedback interaction. It is a theory that makes the juxtaposition of biology and psychology anachronistic, and unifies the two in behavioral science and neuroscience research. (1983e, 250)

As elsewhere, Money was searching for concepts that understood the inevitability of causality but did not reduce causality to a simple one-to-one relationship, in which one effect could be explained by one cause or in which one cause would lead to one effect. Instead, he viewed the phylism as defined by interactionism. In some cases, however, the interaction could be weighted in one direction or the other. This is the process in his study of a much less common syndrome, abuse dwarfism, in which a child's physical development is limited because of physical and/or psychological abuse. In 1976, Money had defined Baron Munchausen syndrome by proxy. In Baron Munchausen syndrome, a person creates a factitious disease as a product of a pathological desire to receive medical attention. In Baron Munchausen syndrome by proxy, a parent creates a disease not in the self but in the child. In a similar example of his search for the resonant term, Money called abuse dwarfism the Kaspar Hauser syndrome, after the mysterious boy whose hidden origins and mystical presence have inspired so many stories (Kitchen 2001). Money discussed the way the paraphilias involved in such syndromes transformed the human:

> Agony becomes ecstasy. The religious zealot, suspended from a wooden frame by fleshhooks through his shoulder muscles experiences not pain, but its metamorphosis into religious ecstasy. The paraphilic masochist, in the same position, experiences sexuerotical ecstasy. These are exemplifications of the principle of opponent-process, experimentally investigated and named by the psychologist Richard Solomon (1980). In abuse dwarfism, when addiction to abuse is manifest, opponent process is at work. Speculatively, opponent

> process may be linked to release within the brain of the brain's own opioids. They transform the pain of an abusive act into a euphoric rush, or high. It is a transformation that forges a bond of appeasement between the abused and the abuser. (1992a, 100)

The surprise here might be in Money's assertion of the erotic base of this relationship between abusive parent and abused child. As in Baron Munchausen syndrome by proxy, in the Kaspar Hauser syndrome, parent and child engage in a mutual addiction that conforms to Money's view of the vandalized lovemap. The intensity of the pair-bond seems to fulfill Money's search for the romantic character of even the most unusual and even debilitating relationship.

The connection between the lovemap and Money's intersex studies was clear from the beginning. Some of his theories were a product of translating an observation from intersex development to those who were not intersex. Thus, the study of Turner syndrome led Money to an early interest in dyslexia. The process was just as often happenstance. In 1987, Money was invited to speak to the twenty-fifth anniversary of the National Institute of Child Health and Human Development (NICHD). Money had been supported by NICHD since its beginning, a fact of which he was always very proud. In his speech, he emphasized many positive results of their support of his longitudinal studies, and one of these was the serendipitous:

> Another serendipity is one that pertained to a few patients who were being followed simply because they had an endocrine diagnosis and I was studying all of the endocrine disorders and their psychology and it turned out by the time they got grown up that they had what used to be called a perversion sexually and is now officially known as a paraphilic status, eroto-sexual paraphilia. And so these ten cases, young adults now, are the first known cases in which there is a prospective record of the juvenile origin and development of a sexual disorder. It's going to be, I hope, very important with regard to society's attitudes towards punishment and imprisonment instead of prevention and medical treatment. And for that reason it is a significant finding particularly in view of the widespread and erroneous popular beliefs and legal beliefs too that these so-called perversions or paraphilias are caused by exposure to pornography, which is absolutely not true. (1987b)

The process and the assumptions are both worthy of note. The process is what might be called a research opportunity. Money knew that the only way of coming to accurate conclusions about the development of gender and sexuality would be through clinical observation of an entire life. The rarity of such an opportunity is obvious. If he could maintain contact with individuals first brought to Johns Hopkins as infants, then he would be able to perform the lifelong observation that was required.

But for such cases to be exemplary of the general population, the individuals needed to be exemplary of the general population. He seldom referred to this assumption, but it is of particular interest given the controversy of so many of his ideas, particularly those associated with his protocol for surgical intervention with intersex infants and with the Reimer case. In these longitudinal studies, he clearly believed that the individuals who were in the clinic "simply because they had an endocrine diagnosis" could be considered to be no different from other human beings, and thus they were viable examples of gender and sexual development. If they developed paraphilias, these would be the same paraphilias as would be experienced by those without an endocrine diagnosis. In this instance, what is so notable is not that he considered these individuals to be psychologically normal but rather that he perceived "normal" to be such a broad category. Regardless of how exceptional was the behaviour, he saw the individual as within the realm covered by his lovemap.

On the one hand, Money was absolutely correct that he was able to create a record of the development of a sexual disorder, but as in all aspects of psychology the "origin" is open to debate. I have examined in detail only one of these cases, and, while my common sense supports Money's assumption of the origin of the paraphilia, I am no more assured than I would be in response to any patient's account of the importance of various events of early childhood. Money might argue that a child's account of a trauma a few years before is more accurate than an adult's account of a childhood trauma, but this is only partially convincing. To then suggest his studies might change the legal system seems a rather extravagant leap of faith. I doubt whether an origin, speculative or proven, would change societal desires for punishment. Like his comment that pedophilia is not linked to pornography, he concluded with yet another attack on the anti-pornography movement – one of Money's favourite targets. In recent history, no responsible person has claimed pornography to be the cause of paraphilia. Rather, the claims have been that pornography precipitates paraphilia or aggravates it, quite different matters that nothing in Money's studies said anything about. But did Money's longitudinal studies enhance our understanding of the way sexualities develop? Absolutely. And did they contribute to prove the value of his lovemap theory? I would say yes.

Money's lovemap was innovative, and his explorations of its possibilities were global. They involved all aspects of gender and sexuality and extended to various cultures of the world. His influence was felt in seemingly unlikely places, such as a long article in the December 2000 issue of the *Atlantic Monthly* on apotemnophilia, the desire to be an elective amputee (Elliott 2000). For most, the obsessive desire to remove a limb is insane. It takes the limits of the sexuerotic even beyond those involved in the Baron Munchausen or Kaspar Hauser syndrome. Typical of the genre, the article took what must appear to many readers to be a strange and even impossibly absurd form of compulsion and examined it in detail to attempt to make some sense of it for the intelligent reader. The article began with the author making a trip to New Zealand to read the papers of John Money, who "published the first modern case history of what he termed 'apotemnophilia'" (73). While Money did not claim that apotemnophilia was the same as the transsexual desire for genital surgery, he recognized the connections, and, in a nonjudgmental fashion, he tried to understand apotemnophilia as part of the sex-gender potential of the human. In the far reaches of psychological disorder appeared John Money and his lovemap.

Money seldom discussed other theorists. Eleanor Maccoby's (1974) biologically based theories of sex difference, for example, would presumably be one of the major points of opposition for the lovemap, but I have found only slight references to her work in Money, most quite late. He acknowledged some contemporaries with whom he more or less agreed, such as professor of psychiatry Robert Stoller, author of many studies of sex and gender and of erotic deviance, but those mentions are few as well. He was even less likely to refer to earlier theorists who had posited similar ideas, such as Austrian physician and Freudian psychologist Wilhelm Stekel. Money especially avoided those who presented these ideas in association with psychoanalysis, which Money came to believe was unscientific and too devoted to spurious searches for causes.

Not that Money was beyond imaginative speculations about origins. One that offered an unusual opposition to Freud's anal and oral stages appeared in an article from 1967:

> The developmental – and evolutionary – relationship of oral and anal functioning to sex has less to do with copulation than parturition and infant grooming. To give birth, mammals must use the

> mouth in proximity to the vagina and must eat the afterbirth. In childcare they lick the young. Even a species as high as the chimpanzee cleans the baby of feces by licking them away. It is in these biological sources that one most sensibly looks for the relationship between the head and the rear parts in sex, whether as rehearsed fragmentarily in infantile development or coordinated with orgasm in the sexual proclivities of adults. (1967b; 1986d, 533)

Money then went on to suggest that toilet training should be seen as a form of animal imprinting, as in litter boxes: "These considerations of imprinting need to be incorporated into the total body of childhood psychosexual theory" (1967b, 534). Money's turn to animal and evolutionary explanations for lovemap behaviour might seem extravagant, but they are apt examples of his willingness to move beyond orthodox understanding, such as that provided in the 1960s by psychoanalysis. Yet the tenor of "sensibly looks" or "considerations of" needs to be recognized as well. Money was interested in exploring causes, but they were not key to the lovemap. The lovemap demonstrated the coordinates and intersections rather than the evolution.

Still, there are many similarities between Money's work and that of psychoanalysts, particularly in the area of the lovemap. Stekel's *Disorders of the Instincts and the Emotions: The Parapathic Disorders*, published in 1924, is but one example. The term *parapathia* existed before Stekel, but he seems to have been the only one who used it consistently, and, according to the *OED*, it is "now rare." Stekel (1924, 15) wrote, "Paraphilia (perversion) frequently shows us nothing more than the well-known picture of parapathia (neurosis). In many cases paraphilia (perversion) is a positive of the parapathia (neurosis)." Thus, for him, paraphilia was a direct reflection of the psychoanalytic disorder. By the end of the book, however, he came to a quite Money-like conclusion:

> The chief feature of our education to-day is that it is an education under anxiety. Fetishism is a self-protective measure against this anxiety. As long as the fiction of a hell and a punishment in the Beyond of an almighty judge in heaven are supported, we can hardly hope for a general alleviation of these paraphilias. (1924, 346)

His final remarks were very suggestive of Money: "Fetishism is a social disease" (347). Vandalized lovemaps anyone?

It is not that I am claiming Money derived his ideas from Stekel. Rather, it seems telling that Money should come up with quite similar ideas with no reference to this famous predecessor. There are others whose ideas are not acknowledged in Money's writing, even in his notes. He seems to have thought of himself as an originator and had limited interest in those who had gone before. He was more at home in a book such as *The Destroying Angel*, attacking the anti-sex crusaders from the past, than in exploring those who might contribute to his vision, such as Stekel. Money was always in pursuit of the great idea and believed he would build it himself. It would be created not from connections with earlier sexologists and contemporaries in the field such as Stoller but from his own thoughts, brought together in one large scientific story.

The problem of the overarching explanation is suggested by the title of a collection of essays on history: *Contesting the Master Narrative* (Cox and Stromquist 1998). I am reminded of a wonderful comment by one of my students: "One thing I learned from this class is that there is no such thing as a universal truth." While many elements of the lovemap are convincing, the whole is too well integrated to seem other than a master narrative. The incorporation of all sexual desire within one schema seems to provide a way of avoiding the exceptionalism that plagues the treatment of sexual deviance, especially when it becomes subject to the law. But the use of terms such as *imprinting* suggests a coherence in human behaviour that is difficult to accept. Regardless of words such as *consideration*, which seem so open, *imprinting* implies a direct correspondence, an inevitability of impact. Given that Money's own sexual and romantic life does not seem to have followed such a simply understandable trajectory, one wonders whether he ever, in even the slightest sense, tried to understand the lovemap of his autobiography. My guess is it would have defeated even his great project of codification.

CHAPTER FOUR

Man and Woman, Boy and Girl

Transsexuality

Man and Woman, Boy and Girl (1972d) was co-authored with Money's postdoctoral student Anke Ehrhardt, who also became a major figure in medical psychology. At its time of publication, Edward Sagarin, long established as the father of the homophile movement, wrote a review that suggested the general reaction:

> When the history of the literature of human sexuality is written, and most of the books that appear from year to year have fallen by the wayside or are remembered only as a footnote in an overdocumented doctoral dissertation, the present book by John Money and Anke Ehrhardt (M & E) will be worthy of a place on the shelf with Krafft-Ebing, Havelock Ellis, Freud, Kinsey, Masters and Johnson, and a very few others. (1973, 271)

From the turn of the millennium, the book has been more criticized than praised, but it remains the first major statement of what Money believed sex and gender were and what they could be. And its title remains an encapsulation of the gender problem, especially for transsexuals.

It is difficult to explain why Money has become such a negative figure for transsexuals. In the 1970s, he would have been seen as second only to endocrinologist and sexologist Harry Benjamin as a fellow traveller, as a non-transsexual who devoted much of his energy to making gender reassignment surgery available to transsexuals. In the case of intersex, the problem is readily understandable: the Money protocol led to surgeries that many intersex individuals believe to be unnecessary and even dangerous. But to the extent that a "Money protocol" existed for transsexuals, it was almost completely positive, coming very close to advice that each individual should be treated exactly as she or he wished. In the 1970s, the Johns Hopkins clinic approved far more transsexuals for surgery than the system could handle. However, *As Nature Made Him* has erased all this work. Colapinto's book ([2000] 2006), the usual citation in transsexual books and websites, tells the story of a man who forced a gender assignment on someone and then did not listen when that person said that the gender assignment was wrong. Colapinto produced a Money who is an example of medical gender rigidity where before he represented medical support, particularly in the establishment of the Johns Hopkins clinic.

A good elucidation of the importance of the transsexual clinic at Johns Hopkins can be found in Joanne J. Meyerowitz's *How Sex Changed: A History of Transsexuality* (2002). The clinic became somewhat of an obsession for Money. His 1995 book, *Gendermaps,* only slightly concerned with transsexuals, ended with the 1966 press release to establish the "Gender Identity Clinic for Transsexuals" at Johns Hopkins as an appendix. In 1977, Money wrote an overview of transsexuality for an exhibit called "By Body Betrayed: A Photographic Exploration of Transsexualism" by Jay Hirsch. There he wrote,

> Early in 1965, Johns Hopkins performed its first complete male-to-female sex-reassignment operation, and later that year formally constituted its Gender Identity Clinic for Transsexualism. The public first became aware of this history on November 22, 1966 when the front page of the *New York Times* carried an official hospital news release. (1977h, 3)

Clearly, the press release always resonated for Money, ten and even thirty years after the clinic was established and long after it was shut down in 1979. It was a text that provided a public imprimatur of the medical treatment of transsexuals, and it offered an archival attack on the closing of the clinic. For another practitioner, such a document might seem minor – a simple statement about a medical event. Some theorists might even think it irrelevant, as the concepts are what mattered rather than the public awareness of them. For Money, however, this press release was part of his conflation of the publication of scientific theory, the extension of medical practice, and the use of publicity to change the world to his way of thinking. The end of the clinic became Money's primary bête noire for the rest of his career, overtaken only in the end by professor of reproductive biology Milton Diamond's critique of Money's work as represented by Colapinto.

The two most extensive endeavours in Money's career were his work on intersex and transsexuality. To an outsider, these might seem easily linked, but there are many ways in which they are very different contexts. While the correct treatment of intersex has been controversial, there has been no doubt as to the reality of the physiology of the condition. Books about intersex have tended to be caught between explaining a very diffuse and complex set of phenomena grouped under the umbrella of intersex and lobbying to support the views of intersex individuals. The modern history of transsexuality probably begins, as Meyerowitz suggests, in the period immediately after the Second World War, with, on the one side, the very public coming out of Christine Jorgensen, the first transsexual whose surgery became a feature in the international press, and, on the other, the growing reputation of Harry Benjamin, the first medical doctor to commit his career to helping transsexuals pursue gender reassignment. While the conflict about intersex is about treatment, the conflict about transsexuality has often been about the nature of the condition. Many have seen it as simply a psychiatric disorder, as it was treated in the *Diagnostic and Statistical Manual of Mental Disorders*, the standard resource for American psychiatry, until 2012. Thus, the pro-transsexual books have tended to have little concern for physiology or even a search for a cause and much more interest in supporting the move to recognize transsexuals to be exactly what they say they are – persons apparently of one sex who should be of the other. This is the argument in very different books such as *Transgender Warriors: Making History from Joan of Arc to RuPaul* (1996) by Leslie Feinberg, *Sex Change, Social Change: Reflections on Identity, Institutions and Imperialism* (2005) by Viviane K. Namaste, and *The Riddle of Gender: Science, Activism and Transgender Rights* (2005) by Deborah Rudacille.

Australian male model Andrej Pejic, who appears in dresses in top fashion shows, has been depicted as one of the new "femimen." The pregnant trans man has become ubiquitous in afternoon television. In another Australian example, Norrie, a born man who went through a sex change operation to become a woman, became officially neuter in 2010. It is difficult to recall the world in which the Johns Hopkins clinic opened, a world in which an American transgendered person might find professional support from Harry Benjamin and probably no one else.

When the clinic was opened, *Time* magazine (1966) published an article that began with an account about Benjamin:

> The longtime champion of these unhappy people is Berlin-born Dr. Harry Benjamin, 81, who swings his practice between Manhattan and San Francisco. It was he who coined the word transsexual, and his new book, *The Transsexual Phenomenon* (Julian Press; $8.50), is already the standard reference on the problem. (52)

The article made no mention of Money or anyone else at Johns Hopkins except a psychiatrist and two surgeons, but it was apparently Money who was the driving force behind the clinic. Both Meyerowitz and Rudacille quote various colleagues at Johns Hopkins who recalled Money pestering them incessantly until they agreed to create the clinic (Meyerowitz 2002, 221; Rudacille 2005, 119). It was also Money who saw a social potential much wider than the small number of patients that could be handled at Johns Hopkins. A book edited by Money and his then student Richard Green, *Transsexualism and Sex Reassignment,* was published by Johns Hopkins in 1969. It included a long chapter by Money and Florence Schwartz on "Public Opinion and Social Issues in Transsexualism" (1969a), primarily about the press release for the clinic.

Money and Schwartz referred to the first sex reassignment at Johns Hopkins, in 1965, and the attempt to keep the information from the popular press. Then a reporter from the *New York Times* contacted Money: "This presented a crisis. Still in its pioneer stage, the gender-identity committee was confronted with the dilemma of whether to release information or remain silent" (1969a, 256). With the success of the careful press release, "the medical profession had carried the public with it and could continue its investigative work under public endorsement, instead of under the pressure of public censure" (256). The article also considered the legal ramifications, as the Johns Hopkins group did not seek justification through finding a precedent that assured them that their clinic was legal but instead "pursued a policy that set a precedent for the law to look to" (258). Part of this was having the "gender identity committee" meet with the university lawyer and a judge who had been involved with the patient earlier. The legal questions were central because those involved wished to treat the surgery as completely a medical choice:

> The philosophical question of freedom of the will and free choice with reference to a patient's right to acquire a sex change was not brought up for specific examination among members of The Johns

> Hopkins Gender Identity Committee. It was simply assumed that, according to the traditions of medicine, the experts would make the decision concerning the diagnosis, suitability, and acceptance of applicants for surgery. (261)

This dispassionate representation is in many ways about a different "freedom of the will," as Money saw the clinic part of the age-old "traditions of medicine" in which he and other "experts" could make the decisions. Medicine was the answer.

Because Money was often operating at society's edges, he was very concerned with how medical assessments could be made when the general populace might be uncomfortable with the issues at hand. In 1977, he wrote:

> Nonjudgmentalism toward the patient as the person manifesting a syndrome is not synonymous with absence of judgment regarding the prognosis and treatment. Nor is the nonjudgmentalism toward the patient synonymous with lack of judgment regarding one's own personal conduct and morality. Nonjudgmentalism toward the patient is professional nonjudgmentalism. It is as essential to one's medical professionalism as is technical competence. (1977j; 1986d, 540)

In the 1969 chapter (with Schwartz), Money provided an assessment that continues to be a reasonable – perhaps the best – argument for gender reassignment surgery:

> The reassignment procedure is justified to both society and the individual on the grounds that, though it is not a cure, conversion surgery constitutes an investigative, ameliorative therapy which helps the patient to live a more contented and productive existence which could not otherwise today be achieved. (1969a, 263)

In a 1986 interview, he presented a less theoretical justification: "you have a whole lot of those patients who will go and get a gun and blow their own genitals off if you don't do it, and I've had several who got knives and cut themselves trying to get rid of their sex organs" (1986a, 36). One example was published in "Erotic Imagery and Self-Castration in Transvestism/Transsexualism" in 1977: "After nearly 30 years of marriage, a 51-year-old man castrated himself in order to fulfill a long-standing fantasy of being a girl" (1977f, 359). Money's response was telling:

> The bizarreness of the act of self-castration and the repulsion that this may evoke suggest that the patient may have qualified as schizophrenic – or at least as psychotically depressed. Self-castration, however, can be, as in our patient, the only behavioral anomaly in an otherwise coherent man. (365)

Money could recognize the coherent behaviour of a transsexual. His ability to provide a rational interpretation of the most extreme actions and to turn that interpretation into a theoretical justification for medical intervention at the fringes of sexuality and gender simply added to his power as a sexual liberationist.

The status Money achieved as the go-to scholar on transsexuality is suggested by the entry he wrote for Grolier's *Encyclopedia Americana* in 1978 (1978h). The editor, Bernard S. Cayne, had already commissioned and received an entry for "Transsexualism" but was troubled by parts of it that seemed at variance with what he was seeing in the popular media. Thus, he contacted the Erickson Educational Foundation, primarily known for its support of research and the dissemination of information on transsexuals. It suggested that Money write the entry. At that time, Money was on the Erickson board, and both his publications and the Johns Hopkins Gender Identity Clinic were financially supported by the foundation. This scenario might seem simply like another example of Money's typical self-promotion and tendency to pursue very actively any source of financial and other support. However, at the time, influential philanthropist and trans man Reed Erickson and his foundation were very much at the forefront of promotion of transsexual surgery. That they should have committed so completely to Money shows the respect he was receiving within the transsexual community.

The extensive confrontations with Money's theories, as exemplified in Janice Raymond's classic anti-transsexual book, *The Transsexual Empire*, present the other side of Money's fame in the world of transsexual surgery. In her book, Raymond ([1979] 1994, 22) stated first that "the Johns Hopkins Clinic has served as a model for others of the same nature." She suggested that many experts had been involved with the clinic, but she named only one: "Of all the persons who have been engaged in this work at Johns Hopkins John Money, now Professor of Medical Psychology and Pediatrics at the Johns Hopkins Hospital, has been the foremost publicist of the transsexual phenomenon" (22). The word *publicist* seems pejorative and snide, but it is accurate. Whatever sexual issue drew Money's attention was immediately and extensively brought to the attention of the public. Money was quite intentionally a proselytizer of "the transsexual phenomenon." If this was, in Raymond's terms, an "empire," then Money was among the most vociferous imperialists.

Money was always interested in expanding the possibilities of studies related to intersex, as in his rather surprising move towards the study of dyslexia in the early 1960s. In his 1963 article on "Factors in the Genesis of Homosexuality" (1963c), Money placed "eonists," those who would now be called transsexuals, in much the same category as various intersex conditions, or those whose gender identity is contradictory to their chromosomal sex. By 1966, however, in a study based on the drawing-a-person test, he began to see more of the complexities:

> Most male eonists are urgently insistent on having a feminine identity, so that one would not have been surprised had they projected their female image into the choice of which sex to draw first. Eonists are also, however, explicitly able to recognize their dilemma and the discrepancy between their sexual morphology and their feeling of feminine identity. Evidently it is this dilemma that may find expression in choosing the sex of a first human figure drawing. In some drawings the dilemma also manifested itself in the basically mannish diathesis beneath the female exterior. Analogous statements may be made for female eonists. (1966a, 161)

The problem thus is not something in the chromosomes or in the hormones, nor is it some kind of psychiatric obsession that deludes the patient. Instead, it is that urgent insistence combined with the explicit ability to recognize "the dilemma." Even today, any clinician who attempts to go beyond treating transsexuals and to understand "the dilemma" meets this nexus. Money's careful articulation recognizes the difference between delusion and presentation of a condition that has an efficacious answer in medical treatment.

There are many examples in the collection edited by Money and Green that show both Money's early insights and how they connect to his other studies – and how he used his position as a psychologist to develop a model of care. In an interview with *Cosmopolitan* in 1990, he recalled one of his central contributions:

> I was the first to publish the idea of a real-life test for people contemplating a sex change: A man would live as a woman, or a woman as a

> man, for at least two years before surgery. When this difficult test is imposed on people, they have a chance to discover whether the sex change is just a screwball fantasy. In the last five years, I haven't heard too much about people who've made a mistake – I think it's because we're very careful now about the two-year screening. (1990b, 111)

From "I was the first" to the suggestion of the alternative as "just a screwball fantasy," this is classic Money in the popular press. He established his own right to speak as the first, the leading, or the most important. Then he acknowledged the view that his practice might seem ludicrous to some readers, and actually validated that view, before asserting that his expertise enabled him to assure the readers that the ludicrous was excluded by medical science.

Again arising from his intersex work, Money explained transsexual understanding through his view of complementary gender identifications, that each person must construct an idea of both the gender of self and the gender of not-self:

> The opposite or negative-valence model becomes a constant reminder of how one should *not* act. This opposite-sex model does exist in the brain as a neurocognitional entity which under specific circumstances, developmental, experimental, or pathological, has the possibility of becoming operational. Transsexualism is perhaps the most intense and dramatic example of the embodiment of this possibility. (1969c, 130)

This is a common-sense view rather than anything that can be coherently tested, but it remains compelling. If we all conform to the two-gender model, then we must know both what to do and what not to do and thus what to be and what not to be. The transsexual is not some biological freak of a female brain in a male body but rather a psychological transformation in which the complementary identities that exist in all brains are firmly and resolutely transposed. The transsexual becomes compelled to identify as what is clearly the "not to be":

> The male transsexual is able to live, work, think and make love as a woman. His female personality is, in part, his conception of those traits and behavior patterns which typically constitute femininity. This assimilation of traits does not create a less genuine or less stable

> female personality. It simply excludes traits such as the urge to fondle the newborn and erotic arousal not by visual and narrative stimuli but by touch, because they are normally outside male experience and comprehension. The male transsexual conforms to the conception of femininity he has assimilated, until by most standards, his personality becomes *her* personality, female, and completely dissociated from male identity. (1969c, 131)

Thus, in 1969, Money offered a clear explanation for the constant complaint, often emphasized by opponents such as Janice Raymond, that trans women are too feminine, that they fulfill all the sexist assumptions as to what a woman should be. To suggest that a trans woman begins with a male understanding of what a female is does not deny the intrinsic worth of her belief that she is a female. Rather, the point is that, unlike for example, a person who is complete androgen insensitive XY, the trans woman enters the world and develops her first gender identity within a male gender role, regardless of how early she develops a belief in her female identity and regardless of how early and completely the female identity and male role come into conflict. The most effeminate male is reared as a boy and begins life within a male gender role. The female gender role is perceived as an external complement. Accordingly, when certain attributes, such as nurturing and a touch response more than a visual one, are not in the foreground in this perception, they will not be in the foreground when the female gender role becomes internalized. To see this gender role as less valid because it is constructed this way and not simply innate is to fail to recognize that all gender roles are shaped socially and environmentally. The female is nurturing and tactile not necessarily because of some innate maternal instinct but possibly because, when she is identified at birth as a female, nurturance and tactility are reinforced in a variety of overt and covert ways, which the person born male who becomes female after the roles are well defined does not experience.

One clear problem with Money's early and often quite definitive descriptions of the transsexual that claimed to offer something akin to a permanent truth is that some truths seem less permanent than others. The chapter titled "Sexual Dimorphism in the Psychology of Male Transsexuals" recorded the reports of childhood experiences by male to female transsexuals: "Labelled 'sissy' by their peers, they much preferred the security of the home and little girl activities" (1969c, 119). On the contrary, Greg Lehne, in a conversation in 2010, suggested that many transsexuals had apparently gender role-appropriate childhoods, although they state that they covertly identified with the other gender at an early age. Even in 1969, however, Money noted the problem of "retrospective" (1969c, 119) information and the inevitability of the transsexual offering a narrative of childhood that convinces the medical interviewer of the essential nature of the opposite gender identity.

The problem of biology is central for transsexuals, in many different ways. While many speculate as to a genetic reason for homosexuality, same sex sexual activity is common throughout the world and throughout history. In other words, the actions that define *homosexual* come very close to being "normal," regardless of whether they constitute an identity or how a society's moral values respond to these actions. On the other hand, while cross-dressing is common in theatre and religious ritual, the assertion that one is simply not the sex of birth but rather the opposite is almost always associated with a specific difference. Many cultures have a third sex, neither male nor female, but in those cultures the prospective members of that third sex are usually self-selected by the way they move in the world. The community observes that they differ, and they are allowed – or in many instances forced – to enter that third sex (Roscoe 2000). In some cases, such as the *hijra* in India, a number claim some intersex characteristics, but even the hijra usually have no obvious physiological difference. Yet, as the many studies of the hijra show, outsiders are compelled to find one (Reddy 2005). There is an undercurrent of this imperative in the questions Money asked male to female transsexuals about their genitals:

> The male transsexuals invariably chose the receptor role in sexual relations. Direct stimulation of the penis was unpleasant and undesired; sexual receptiveness was not always accompanied by erection of the penis or other evidence of responsivity of erotic tissue. Whether the dysfunction of the male genitals may be attributed to psychodynamic factors and/or treatment with estrogen, in no case was there regret at playing a receptor role only. On the contrary, there was delight in it as a symbol of femininity. (1969c, 121)

Money, like most sexologists of his generation, had quite a bit of faith in the penis. It was not that they believed that a man's sex was simply a product of his penis but that the gender identity of the male was so closely tied to the penis that its activity was central to his sense of self. Thus, the transvestite who used women's clothing to stimulate an erection and achieve orgasm fit neatly into Money's paradigm of paraphilia. The transsexual was something quite different:

> There is a dissociative quality to the male transsexual's disengagement of genitopelvic erotic functioning and his engagement of feminine imagery-content, while having sexual relations. As in hysterical dissociations, such as conversion blindness or conversion paralysis, neural and neuropsychologic functioning are altered so that there appears to be no regret at the dysfunction or loss, namely, insensitivity in the normal center of erotic stimulation, the penis. (1969c, 125)

Money noted the similarity to "hysterical dissociations," but, as in the observation about self-castration, this similarity is of function, not of etiology or of appropriate medical response. Still, it is revealing that Money presented "the normal center of erotic stimulation, the penis." Arguably, he should have said "normal male center" or even "normal male non-transsexual center," but just the word *normal* suggests a rather constrained view of male sexuality in general. Money noted that an adult male who loses penile function tends to continue to feel the penis in his dreams and often senses a genitopelvic reaction even when there is no penis at all, but this does not seem to happen for post-operative transsexuals: "The absence of any penile sensation following amputation may be the result of previous dissociation of function" (126). The "normal" male without a penis has penis dreams, whereas the transsexual erases the penis. While this might seem limiting in many ways, particularly in assuming that maleness is defined by the penis, it is also affirming of the trans woman. Raymond ([1979] 1994) and others might see her as a deluded male, but, if the biological male is defined by his penis and the trans woman, regardless of being formerly a biological male, does not have even an absent penis, then Raymond must be wrong.

Money's incessantly inquiring mind led him to a variety of observations and conclusions, including one that would eventually suggest a basic difference between the male heterosexual transvestite and the male to female transsexual. In 1967, he wrote the following:

> I have been struck by a peculiar twist to the imagery in masturbation or coital fantasies of male transvestites or transsexuals whose bedroom sexual activity is heterosexual (with a woman). By this twist of imagery, the affected male sees himself in his mind's eye as a female with feminine body shape and organs, undressed or wearing articles of female clothing. In this imagery, the male organs and physique are transposed to the female partner, though perhaps with some misgiving for such an act of debasement and disrespect. To the best of my

> knowledge, this double inversion of imagery is specific to the transvestite or transsexualist. The extremely effeminate homosexual might ascribe to himself the image of the female, but he will be in bed with a male whose image he will not change, except maybe to enhance its masculine attributes. (1967a, 142)

At this point, the distinction between transsexual and transvestite was not clear, but, as he began to recognize the importance of this masturbatory transference, he observed that, while both were engaged in the same process in this scenario, the transsexual would be happy to reach the stage noted above – the absent penis – but, for the transvestite, the erection of the self was central to the process. In other words, the cross-dressing heterosexual was creating a dream transference to the female partner in support of his own penis. The homosexual would act as a female to enhance both his own penis and his partner's emphatically male penis. The transsexual was seeking the end of the penis yet remained in a world emphatically heterosexual, a heterosexuality that he, as a male, accepted and that his vision of himself as a female ironically reinforced in the transference of complementarity.

In general, Money was quite sensitive to the needs of his patients, but he was always compelled by science. Thus, his medical position on transsexuals was completely in favour of them making their own decisions about surgery, after the requisite real-life test, but his scientific assessment of them searched for something much more objective and defined. Money's experience of intersex conditions led him to look constantly for hormonal explanations for transsexuality: "Allowing that vulnerability to error in the achievement of normal psychosexual identity may be created by early hormone exposure, then it is possible that some of the uncanny clinical similarities of transsexuals may be due to the consistent nature of an as yet unknown fetal hormonal event" (1969c, 128). He considered the various studies of the way hormonal treatment can change rat behaviour, but I have found no evidence that he saw that these "uncanny clinical similarities" also could reflect narratology. As Jay Prosser (1998, 107) suggests, long before transsexuals became a part of the popular imagination, "the intelligible transsexual story is always already understood, not bizarre and foreign but familiar, anticipated and – quite crucially – named." The person who recognizes him- or herself to be transsexual recognizes the need for a specific narrative that can justify that recognition. This is a somewhat strange absence in that Money was so aware of the power of language and the power of socialization in how a gender role was configured. The similarities in gender performances, often in ways that have no apparent connection to hormonal influence, could at least partly reflect the narrative necessity felt by transsexuals who very early recognize that the story and rhetoric of their processes must conform to something identifiable as truth.

Money's sensitivity to the sense of self in the transsexual – that the transsexual decides what is necessary – did not seem to extend to understanding that, if the transsexual is convinced of her gender identity, then it is likely that the autobiographical narrative will be shaped according to that identity and that both the conscious and the unconscious selves will conform. The narrative necessity offers an alternative explanation to his reaction to apparent memory lapses in transsexual patients in a study in the mid-1960s:

> Yet another point for the record, apropos of the comment of Worden and Marsh (1955) regarding memory in transvestites: there were frequent instances of memory and recall failures in the children I have studied, with regard both to manifestations of femininity and to other aspects and problems of their lives. These failures included oversight, selective editing, denial, suppression and, apparently, genuine repression and forgetting. One is tempted to speak of an amnesic or dissociative talent as a special characteristic of the mental functioning of these youngsters which permits, with unusual ease, the operation of hysterical, dissociative personality mechanisms and defenses as in the dissociation of masculine phenotype of the body from feminine interest and activities. (1967a, 142)

The concept of dissociation is of particular interest in transsexuality. If, as male to female transsexuals often asserted long before Money's work, the brain is simply female in a body that is male, then it is not dissociation but a disorder of the sort any human might have with a conflict between the body and some foreign component within it. If the situation is instead dissociation, then it likely might be connected to many other dissociations. But while Money had diagnosed dissociation, he also decided that the most efficacious treatment of the dissociation was to correct the body to suit the dissociation and thus, apparently, erase the dissociation. This then, in at least some examples, might erase the problem of the amnesia because, rather than a classic amnesia, it is an erasure of memories that are no longer appropriate for the corrected body. The same could be the case for apparently invented memories. The trans woman has lost the boy memories and instead has girl memories. Moreover, this can explain many gaps that on the surface do not seem to be about gender.

The human mind creates dissociation in the aid of association. Thus, the present self requires a past self that is not dissociated from the present and will dissociate him- or herself from memories that create dissociative contradictions. In other words, the transsexual is not that different from the rest of us.

Money believed the childhood of the subject in any sexological study offered the best answers to etiology, but his was not a psychoanalytic process. Money's usual opposition to psychoanalysis was made specific in a review he wrote of Robert Stoller's book *Sex and Gender* (1968):

> Stoller recognized the importance of clinging and body contact in his investigative psychoanalytic studies of mothers and their young transsexual sons. Insofar as I can judge from his book, however, he became snared by the metaphysic of his analytic training and slipped, unwittingly, into the position of laying all the responsibility on the parents, especially the mother. In this way, the offspring becomes a cipher, who could be replaced by any other cipher. (1970c, 226)

Of course, as Meyerowitz, Green, and others suggest, Stoller was more or less Money's chief competition in theories of transsexuality at the time. And Money never dealt well with competition. Still, Money's position was deeply felt: he feared the tendency to identify all aspects of the psyche, particularly of psyches identified as abnormal, as products of parental influence. This was a particular problem for transsexuals. If the transsexual was an effect of parental dysfunction, then the answer was lengthy psychotherapy rather than the affirmative transformation offered by gender reassignment surgery. Instead, Money wished to accept the identity presented by the transsexual but to trace the development of that identity through emphasizing the various points at which any intervention – hormonal, physiological, psychological – might have the greatest effect on the future adult. Thus, his understanding of transsexuals really began with the analysis of effeminate boys that he did with Green, with the first results published in 1960. The study arose from intersex work because Money and others assumed there might be a hormonal or chromosomal explanation. When I asked the endocrinologist involved in the study, Claude Migeon, whether there was any suggestion of chromosomal deviation, he replied, "All transsexuals are normal." Migeon later said to me, "I should have said 'most transgenders have a normal karyotype – I have learned that nothing is absolute.'" Still, his original statement seems applicable if one sticks to biological factors that are medically discernible. There might be a few exceptions, but "transsexuals are normal."

Money and Green found no suggestion of physiological difference. They saw some possible causal factors in parenting, but these were not sufficient to establish a clear etiology. The conclusive results tended to be what common sense might have expected, such as the way the behaviours of the effeminate boys conformed to female behaviour:

> It is generally considered, both in popular and scientific belief, that females are verbally superior to males whose strength lies in mechanics and mathematics. There is a certain consistency, then, in the fact that male eonists and prepubertal effeminate boys, in whom feminine identification is prominent, have a tendency toward verbal superiority in the Wechsler test – a tendency that is more clearly evident in the special factor scores than in the less sensitive comparison of Verbal and Performance IQs. (1967e, 452)

While Money identified the lack of certain aspects of feminine behaviour in the male to female transsexual, as in his observations on nurturing and touch, this did not mean that the trans woman's identity could be subsumed in a pseudo-girlish performance of the sort attacked by Raymond. The Wechsler test suggested that this female identity is more full and complex than that.

Yet part of this complexity was also a discovery that feminine behaviour in boys, no matter how it was constituted, did not necessarily produce a female identity in the adult. This became a central point in Money's various comments on the value of his longitudinal studies, as in his speech on the twenty-fifth anniversary of the National Institute of Child Health and Human Development: "I undertook beginning in the late fifties a prospective twenty-year outcome study of the juvenile gender status of little boys who were behaviorally very girlish and this effeminacy proved to be a precursor of a homosexual gender status in adulthood in all the cases that could be found for follow-up, but it did not prove to be a precursor of either transvestism or transsexualism" (1987b). Given that the idea of transgender children has become part of popular culture, this is a particularly notable observation. In 2008, the *Atlantic Monthly* published "A Boy's Life," in which Hanna Rosin examined the case of Brandon, an eight-year-old boy. Interestingly, Rosin barely mentioned Money's work on transsexuals; instead, she referred at length to the Reimer case, which she set up this way: "In 1967,

Dr. John Money launched an experiment that he thought might confirm some of the more radical ideas emerging in feminist thought" (2008). Not surprisingly, she then turned to Colapinto and Diamond and concluded, "Behaviors are fundamental unless we are chemically altered." She considered the extreme treatment reactions available to transgender children, from Dr. Norman Spack, who gives them hormone blockers, to Dr. Kenneth Zucker, who uses therapy to convince the children to accept their birth sex. So where does this leave Brandon? According to Rosin, although Brandon is living as Bridget, he or she is not happy. Then, of course, there is the mother of Bridget's best friend, Abby, who refuses to let Abby see her because "God doesn't make mistakes." Transsexual or homosexual, the religious right will not accept Brandon or Bridget. Money would have hated the blind discrimination of the outraged Christian moralist, but, in many ways, this is a red herring. A more important point of discussion is whether Money's observation in his longitudinal study presents a permanent fact of gender development or is perhaps a product of very temporal social forces. The gap here is between possible outcomes given the force of variable narratives. There was no article in the *Atlantic Monthly* on the subjects of the study by Money and Green. When they became adults in the third quarter of the twentieth century, they proved not to be trans women, but will the same be said of the effeminate boys who become adults in the second quarter of the twenty-first century? Especially given that popular culture continues to have a rather simplistic view of the problem, as in the title of a *Maclean's* magazine article on the topic: "Boys Will Be Girls" (Gulli 2014).

As in the case of the paraphiliac and the lovemap, Money attempted to place the transsexual within larger schema of sexological assessment. The issue of the self-diagnosis was one aspect. In one unpublished article from 1981, Money offered a chatty and accepting response to the process:

> Medicine has a millennial history of complaints that are self-diagnosed – headache, for example, and many other aches and pains. For a similarly long period of history, self-diagnosis has been related to self-prognosis, and self-therapy, as in self-medication (with, e.g., alcohol), self-incision and even self-destruction. It is but a short step in from self-therapy to enlisting the assistance of an expert, as, for example, in cosmetic alteration of the body by means of tattoo, scarification, and surgical alteration of the genitals. From folk practice to modern cosmetic surgery is, conceptually, but another small step. (1981a)

Money noted that the appropriate medical response is not to dismiss self-diagnosis but rather to assess it as useful or not in terms of treatment. His reference to "modern cosmetic surgery" suggested that treatment can be not just a product of progress in medical knowledge and method but also social change, in which the correct treatment in a present, perhaps more liberal, era might just be different. Some thirteen years later, in "Body-Image Syndromes in Sexology" (1994), he presented a somewhat less accommodating response to those who self-diagnosed. He listed some of the attributes he had encountered that suggested an obsessive fixation on a need to modify the body:

- a history of multiple prior evaluations
- self-referral by telephone or letter
- expansive or fancy handwriting on the envelope
- a presumption that it will be an honor to have such a rare case for either research or treatment
- money is said to be no object even though none is available to pay bills
- patient assigns to the doctor the role of technical assistant to himself, who is the expert on the case. (1994, 46)

The tenor here, however, seems at least partly a response to Money's long-standing public status as an expert. Throughout his career, Money seemed a likely resource for those who perceived themselves in some sense at the extreme edge of gender or sexual possibilities, and his papers are full of letters from such persons, including a number from inmates of penitentiaries. This might have led to understandable irritation at the arrogance that the self-diagnosed had at times about the uniqueness of his or her case. On the other hand, one could argue that Money should have been more sympathetic to the fact that each of us is an expert on his or her own problems. Most of these patients had many years of introspection in a society where they feared that their problem would be discovered by the world of the "normal." One can only guess how rare they felt.

"Body-Image Syndromes in Sexology" is primarily concerned with whether or not problems of body image should be considered to be pathology. The article notes how much medical treatment of what might be called a self-diagnosis of body image disorder conforms to certain processes, regardless of how simple and accepted or how complex and seemingly outlandish the disorder. Thus, rhinoplasty probably is not a procedure that should be seen to represent a body image pathology, yet it follows a pattern that appears in conditions generally regarded as pathologies: "three principles of body image syndromes: 'Realignment and Enhancement,' 'Obliteration and Relinquishment,' 'Augmentation and Amplification'" (1994, 34). Money's division of complex issues into factors more recognizable as structural and thus more easily theorized was often reductive, but the process also showed the similarities between the "normal" and the "abnormal." Here he defined the central desires of someone with body image issues and the actions that can be taken to address them:

> In the case of the male-to-female transsexual, the body image of the face is feminine and is discordant with that of the actual physiognomy ... The principle of alteration according to which congruence between the femininity of the facial body-image and the actual physiognomy is effected is realignment and enhancement. (1994, 34)

Thus, while rhinoplasty and sex reassignment might seem two completely different categories, Money showed that they have similar problems of perception for the patient, between the image as it is and the image as it should be, and similar procedures – surgical altering of the body – are chosen to address that perception.

Money's sensitivity to the most extreme cases and his ability to see the connections to the range of sexual and gender issues were visible in an article he wrote titled "Three Cases of Genital Self-Surgery and Their Relationship to Transsexualism" (1976f). As Money noted, "In all three cases, the patients relied on a reassurance of professional nonjudgmentalism – of not being stigmatized – before they could be open about what they had done to themselves" (293). The first case in this study seems to be that of the artist Forrest Bess, who first wrote to Money in 1962. Money's attempt to thread the difference between different possibilities is clear in the discussion with which he concludes the article:

> The transsexual compulsion typically manifests itself as an idée fixe rather than a paranoid delusion. That is to say, whereas the transsexual patient does not yield ground to an opponent on the issue of his own feeling about being a woman, he readily concedes that by other criteria he is not one. Only in Case 3, in the present series, did the gender ideology qualify as an idée fixe. In the other two cases it became a paranoid proposition, embedded in a delusional system incapable of being consensually validated. (1976f, 293)

For Money, the idée fixe could be its own justification. In the case of the transsexual, the idea could not be changed, but the lack of pathologies in other aspects of the transsexual's life made Money question whether any attempt should be made to change it. Instead, the "fixe" should justify making the body to conform to the "idée." The paranoid proposition is a very different category: it rejects the difference between the self-perception and that of the outside world. The idée fixe recognizes the difference but will take any possible course to erase it. One of the most interesting aspects of Money's observations, however, is not such a psychiatric assessment but rather his dispassionate response to the surgery:

> Self-demand surgery of a nondiseased organ falls into the general category of cosmetic surgery. It is now an accepted part of the theory of plastic surgery that cosmetic surgery may be psychically rehabilitative. It is on this theoretical basis that sex-reassignment surgery is undertaken for transsexualism. In the present three instances, there is no tradition on which to draw in order to prognosticate the possible rehabilitative effects of either the self-performed surgery, or in Cases 1 and 2, of the surgery further requested. It does appear, however, that the self-performed surgery did not result in a deteriorative state, psychiatrically. For the foreseeable future it is unlikely that there will be an answer to the issue of postsurgical outcome, for in the present climate of the rights of clinical investigators, informed consent, and malpractice liability, it is unlikely that surgeons will feel free to cooperate with psychiatrists in order to obtain an answer. Meantime, affected patients will continue to do their own surgery in secret, and the medical profession will subsequently be called upon to do what it can amelioratively and supportively. (1976f, 294)

Money's comments seemed less about etiology or any other theory than about a basic medical question, the best outcome for the patient. He recognized that society was simply not ready to accept the conditions in which such a judgment could be freely made. Instead of the surgeon treating the patient according to the best outcome, and to the devastating effects likely in the absence of surgical treatment, the surgeon and hospital will instead respond to the dangers of "malpractice liability." It is interesting, moreover, that he also included the "rights of clinical investigators." This is a strange phrase, as it is difficult to see what these might be. Today, when studies involving human subjects are very carefully regulated, the issue seems to be much more "the responsibilities of clinical investigators" or even "the severe restrictions on the actions of clinical investigators." Money still had hopes for his rights.

In my own discussions of Money's theories with friends and colleagues, the topic of apotemnophilia causes the most extreme reaction. The most suicidal forms of paraphilia seem more understandable to most than the desire to amputate a healthy limb. For Money, it was one more of the vast array of paraphilias, but he also recognized that it connected to the larger issues of body image and transsexuality. His comments on it in his article "Apotemnophilia: Two Cases of Self-Demand Amputation as a Paraphilia" (1977) demonstrated his careful logic at its best, so dependent on avoiding judgmentalism:

> The idea of self-amputation meets so seldom with consensual acceptance by other people that it might readily be stigmatized as a paranoid delusion. It is, in fact, an idée fixe rather than a delusion. Judging by the two cases of this report, the apotemnophiliac, unlike the paranoic, recognizes that other people do not accept his own idea concerning self-amputation. One may think that apotemnophilia is synonymous with masochism. In the two cases reported, there was no history of the erotization of pain itself, but only of the healed, amputated stump. Thus, apotemnophilia, because it involves injury and pain, bears a peripheral relationship to masochism, but is not identical with it. (1977a, 124)

The reference to the idée fixe provides another link to transsexuality, but the link is not just about an erotic process of removal. Money saw the idée fixe as definitively different from delusion. Whereas the general public and the vast majority of clinicians see apotemnophilia as a mental disorder, Money categorized it as a singular "idée fixe," not a symptom of insanity but an individual idea that could not become unfixed but could be medically addressed by conforming the body to the fixation, regardless of how difficult it is for society to accept this kind of surgery. Money was very careful to assess the subjects' relation to pain, or masochism, and also to the possibility that they just desired surgery, as in Baron Munchausen's syndrome. Instead, Money became convinced that the desire was solely for the amputation. As an idée fixe of body modification, it could be linked to the transsexual: "Though neither wanted an alteration of his gender status, the two patients reported perceived a relationship between amputation and transsexualism in the sense that both involve self-demand surgical alteration of the body" (1977a, 124).

As he listened to their stories, he realized that the similarity went beyond surgical alteration: "Both patients cited a nonerotic imagery of masculine overachievement as providing an erotic turn-on in their amputee fantasies. The over-achievement imagery primarily consisted of amputees overcoming the adversity of a handicap" (125). Not only was there an erotic component in the desire for amputation, but also there was a specific gender transformation. Thus, apotemnophilia is a logical part of Money's lovemap. Since Money's study, there have been a small number of female examples, but, at least in the literature I have read, there has been no consideration of whether there is a gender need in their desire for amputation.

Nonetheless, once again, Money's primary concern was the patient:

> The present status of informed consent in medicine with respect to investigative procedures and the threat of malpractice charges makes it unlikely that there will be an early answer to the question of whether self-demand amputation is an effective form of therapy in apotemnophilia or not. The answer will have to come from patients who have engineered an amputation for themselves and then are generous enough to volunteer themselves for post-surgical study. (1977a, 125)

This is a very different Money from the one depicted by Colapinto. Money took nonjudgmentalism to a point where few would go. He might be seen by many to be an enabler of self-destructive insanity. Yet one can see here a suggestion of why Money never faltered in his commitment to transsexuals. While he was consistently interested in their etiology and in any possible explanations of any aspects of their conditions, he was ultimately convinced that fulfilling their demands for surgery was the most efficacious form of treatment. And there was no question he was the person to enable this treatment.

In 1971, Money published an article in the *Journal of the American Society of Psychosomatic Dentistry and Medicine* in which he suggested that sex reassignment could be a valid choice both psychologically and physiologically:

> In transsexualism, sex-reassignment therapy not only endorses the organism's own attempt at self-healing, but also furthers it by the administration of hormones and performance of surgery. Except for those already familiar with intersexuality and related disorders, such therapy represents a radical departure from tradition. It is small

> wonder, therefore, that the legalistic mind, trained to rely on precedent, should be hesitant to legitimate the new procedure. Eventually, however, the law catches up with history. (1971e, 26)

The law and society were hesitant, but Money's history of working with intersexuality made him see that the treatment was not "a radical departure from tradition." Instead, it was a development from a radical tradition. Money's ultimate goal was not to look back to the precedent of what the law had allowed before; his goal was to look forward to what the person recognizes as the means of "self-healing."

Money's view of what could be done as an agent of self-help eventually became too radical for Johns Hopkins. Bullough, in his "The Contributions of John Money: A Personal View," gave this account of the closing of the gender reassignment clinic:

> In 1975, Paul McHugh, an opponent of transsexual surgery, became chairman of the psychiatry department at Johns Hopkins, and he was determined to end the practice, comparing it to the lobotomies of an earlier generation of psychiatry. John Meyer, the then-head of the clinic and no friend of Money's, agreed and closed it. Since Meyer allegedly did so on the basis of his own poorly conceived study – which had serious methodological flaws including a poor response rate, a lack of any real scale for post-operative adjustment, and numerous other problems – it seems to an outside viewer that the decision was a political one, perhaps aimed at Money by some of his academic rivals (Fleming, Steinman, and Bocknek 1980; Meyer and Reter 1979). (Bullough 2003, 234)

Rudacille (2005) makes McHugh's conservative views very clear in her book. He seems to have been unlikely to accept Money's theories on any topic. Yet the study by Fleming et al. is respected, coherent, and consistently damning. It leaves little doubt that the value of the Meyer study was only in its result. Johns Hopkins wanted a reason to close the clinic, and Meyer's questionable study provided it. As so many have often asserted but everyone, including Money, often seemed to forget, surgical units are ruled by the surgeon. In 1979, the surgeon then in charge of the sex change operations, Howard W. Jones, reached compulsory retirement age. The surgeon was gone, and the psychologist would not be allowed to continue the clinic.

It has always been difficult to assess the success of transsexual surgery. The usual evaluative criterion is the happiness of the patient; and, understandably, most patients, having received the thing for which they had pleaded, claim to be happy. Yet the popular press enjoys presenting the opposite whenever possible. In one typical example, the Australian *Sydney Morning Herald* published an article in 2009 with the title "Gender Setters – When Doctors Play God." Much of the article is about two case studies, but it also includes some accounts of medical opinion:

> Doctors from London's Portman Clinic say they see many patients who feel trapped in "no-man's land" after surgery. Reviews of the Melbourne clinic found psychotherapy was rarely offered. While a patient would require a diagnosis as a "true transsexual" from two psychiatrists before surgery, both opinions were from inside the clinic – one that operated under the ethos that surgery was the only cure. (Stark 2009)

The reference to "no-man's land" is witty, but it is also dismissive, with no sensitivity to the possible validity of the transsexual's demand for gender reassignment. This article also uses the decision of Johns Hopkins to close its clinic thirty years before as a convincing example. The article notes the opposing point of view expressed by most psychiatrists and transsexuals – that failures are rare and unusual, and most transsexuals feel fulfilled and complete after surgery – but these opinions are dismissed.

The opposition to the opposition is ubiquitous – transsexuals who attest to the absolute necessity of surgery. One example, which appeared in the *British Journal of Sexual Medicine* soon after the closure of the Johns Hopkins clinic, was from a British trans man named Nicholas Mason, who reacted to Meyer's study. Both the introduction and Mason's comments referred to him as "Myer," rather surprisingly as Money's response printed on the same page used "Meyer":

> If a mere layman may dare to question the theory of Dr. Myer, one suspects that he may be putting all transsexuals into the same class. Some are undoubtedly disturbed, but then so are some diabetics. One would not infer from that that all diabetics are unhappy. After treatment these transsexuals would still be unhappy and prove Dr. Myer's point. There may be another group, undoubtedly psychotic,

> whose illness manifests itself as a desire to change sex. The rest of us may appear depressed but after treatment this exogenous depression is removed and we are able to live full and happy lives. For this I can never adequately thank all those who have made it possible. (Mason 1981, 61)

One might expect Money to respond positively to such a supportive comment by a patient, but instead Money began by confronting Mason:

> Mr. Mason is in error. Psychiatrist John Meyer did not claim that transsexuals are no happier after surgical treatment than before. Quite to the contrary, he specifically excluded personal happiness from his psychiatric reckonings, though the 15 postoperative transsexuals in his study were, in fact, happy with their new status. None of the 25 in the preoperative group had given up on sex reassignment. (1981e, 62)

Money never let a supporter get away with support that seemed other than what Money deemed appropriate.

That first *Time* magazine article (1966) about the opening of the clinic stated, "Last week's announcement from Johns Hopkins marked the first time that a prestigious medical center has risked its reputation to give organized help to transsexuals." When the clinic closed, *Time* (1973) again had a comment:

> Last week the surgery came under a surprise attack. At Johns Hopkins Hospital in Baltimore, where about 50 sex-change operations have been performed since 1966, doctors announced they were abandoning the surgery for all but hermaphrodites (those born with male and female organs). Reason: a new study finds no difference in long-term adjustment between transsexuals who go under the scalpel and those who do not. (34)

One, of course, cannot expect the media to do more than report what the medical scientists do, but it must be interesting that *Time* should use this tone to suggest that what Johns Hopkins University Hospital gave it could also take away.

Many continue to search for a biological explanation for transsexuality. The most recent example that I have seen was in 2008: "In the largest ever genetic study of transsexuals, Australian researchers have discovered a DNA variation linked to male-to-female transsexualism" (Smith 2008). I have found no follow-up to suggest that this study has ultimately proved any more definitive than any other claims of biological explanations. Thus, the situation remains as it was when Money first entered the field: the diagnosis is made by the self, and the medical procedure is only to fulfill the demand of the self, based on whatever grounds the medical personnel or the state wish to institute. Money sought to answer that demand, but Johns Hopkins did not.

Of course, as noted throughout this book, Money had no doubt as to the absolute necessity of a clear gender identity of either male or female. While he himself recognized that many humans biologically straddled that divide, and many crossed it at different times in his life, he believed that the self needed a clear gender identity to function. This seems somewhat surprising given that his own dissertation depicted a number of people who might have appeared to be one gender or another to the outside world but felt themselves to be straddlers. I have no idea what Money would have made of someone such as the Australian Norrie or, for that matter, the protagonist of Leslie Feinberg's *Stone Butch Blues* (1993), who becomes a man only to decide to move back and live life as an out transgendered person. Feinberg has done much to make the history of this possibility more well known, and today anyone in the community of sexual diversities knows people who choose to be not male or female but transgendered. Whether or not the number of such people will grow or whether they are exceptions that will prove to be only that, who can say? However, at present, there are many people who are taking "Man and Woman" in a direction Money would have believed unlikely to be functional. And according to them, they are quite functional and happy. They are succeeding at self-healing.

CHAPTER FIVE

GAY, STRAIGHT, AND IN-BETWEEN

Homosexuality

In his early years, Money seems to have considered himself heterosexual, but after his brief marriage failed he was very much a part of gay culture and referred in his diary to a few individuals as "the boyfriend." When Money "came out" to the administration at Johns Hopkins in the late 1960s, the reaction was so negative that he became much less open about his sexual orientation. As Richard Green recalled in conversation, Money advised that continuing research on homosexual topics would be career suicide for Green. Yet there is little to suggest that Money's personal relationships with women reflected some attempt at staying in the closet. Later in life, Money had relationships with women that were romantic and sexual but never led to cohabitation. He also had casual sex with women and participated in different types of sex parties. He often served as an expert witness in trials on issues concerning homosexuality, and during one of these trials the prosecution asked him if he was homosexual, and he replied, "No." One of his postdoctoral fellows, Tom Mazur, asked him how he could say that, and he replied, "Because I am bisexual." So his concerns about homosexuality were a personal matter, but his writings never reflected that personal investment. Yet this is true of his work on all subjects: all his books maintained a detached tone, with a decided air of the objective scientist. Money was an ardent publisher, and he decided what would fit in any publication, at times producing books that seem somewhat arbitrarily assembled. Thus, *Gay, Straight, and In-Between* (1988a), with the, excuse the expression, rather straightforward title, in[illegible] that seem at best vaguely related to the topic. As a pediatric [illegible] ologist, his concerns for the possible relevance of intersex c[illegible] seem appropriate, but he also strayed in a variety of other dir[illegible] ing transgender and all of the possible paraphilias. Perhaps he [illegible] to overcome the simplicity of the gay/straight divide, to present sexual orientation as one more of the many aspects of gender and sexuality.

Money had long realized that society's division of the world into male and female did not reflect the expansive continuum of both gender and biological sex. In 1963, he published an article titled "Factors in the Genesis of Homosexuality " (1963c), which began, "Hermaphroditism and related disorders are anomalies that, because of the incongruences inherent in them, raise some puzzling semantic technicalities as to the very definition of homosexuality" (19). Knowledge of intersex had an inevitable effect on beliefs about sexual orientation. Destabilizing gender made it difficult to divide the world into gay and straight. Similarly, homophobic religious doctrines and laws become inapplicable if it is difficult to decide when a sex act is between members of the same sex. Money noted that societal concern for the medical needs of intersex persons can change very quickly when those needs seem to lead to homosexuality:

> No infringement has aroused more fanatical vindictiveness than that which pertains to the morphologic sex of the partner in sexuoerotical practices, namely, when the partner of a male is transposed from female to male so that two males are together, and similarly for two females together. This is the transposition for which the modern term "homosexuality" was coined by K. M. Benkert, also known as Kertbeny, in 1869 (Kennedy 1988; Herzer 1986), although the phenomenon had been on record since antiquity. (1999b, 191)

As so often in Money, a simple process is given an unusual twist through language. *Transposition,* a word more common in music, makes homosexuality simply sex in a different key, with no negative connotations. Money often noted a similar transposition among transsexuals, who have been regarded by various medical boards as unfit for surgery if they intend to continue with the same sexual object, in effect turning from a heterosexual into a homosexual. Money recognized that many negative responses to gender instability could be an extension of society's deep-seated fear of homosexuality. This is reflected in the recent tendency in some cultures, such as Iran, to

[ac]cept transsexuals as diseased persons in need of surgery and yet to reject homosexuals as criminals in need of punishment. Money the pianist would not have been surprised at this illogical failure of modulation.

Money believed that some of the spots on the intersex continuum might provide some insights into what controls the choice of sex objects by all humans. In 1970, Money published a paper called "Sexual Dimorphism and Homosexual Gender Identity." There he stated,

> Errors of differentiation, notably in both clinical and experimentally induced hermaphroditism, are important to the theory of homosexuality. In some instances these errors result in what is, in effect, homosexuality by experiment or by experiment of nature. (1970g, 428).

In other words, someone who has complete androgen insensitivity syndrome, and is thus a woman who is XY, is chromosomally homosexual if she is in a heterosexual relationship with a man. For Money, as for Albert Ellis before and many other theorists after, the intersex individual provided a biological problem that might offer a biological answer to sexual diversity.

Not that it would ever be a simple answer. Money noted in an article he wrote with Richard Green in 1961 that "Effeminate boys (and men) have normal hormone functioning. It is also known that effeminate boys have normal sex chromosomes" (1961c, 2). In a follow-up in 1979, he observed that "each boy had developed a conviction that he should change into a girl, and that he should be able to do so by somehow or other losing his penis" (1979b, 38), and yet they turned out not to be transgendered, as one might have expected, but rather homosexuals with no cross-dressing and no "inadequacy of genital functioning" (39). In other words, while there seemed a clear relationship between childhood "discordant gender identity/role" and adult sexual orientation, it was not a simple continuity from childhood discordant gender identity to adult discordant gender identity.

Money found a similar failure in the search for a hormonal answer to homosexuality. In an article titled "Components of Eroticism in Man" (1961a), he asserted the importance of hormones in the sex drive and yet the failure to differentiate either between male and female or between gay and straight:

> Failure to detect an hormonal anomaly in homosexuals has been paralleled by a failure to effect a cure with hormonal treatment. Androgen does not masculinize the erotic inclination of a male

> homosexual. It either leaves his eroticism unchanged or else intensifies it in the homosexual direction.
>
> The obverse of a lack of abnormal hormonal findings in homosexuality is the lack of increased incidence of homosexuality in sex-endocrine dysfunction. Estrogen levels are elevated in males with gynecomastia sufficiently to cause prominent breast feminization. There is no corresponding feminization of the personality. Androgen levels are elevated in females with hyperactive adrenals or, more rarely, a virilizing tumor. There is no corresponding lesbian virilization of the personality. (246)

Ever the student of both sexuality and gender, Money found here, in a very early study, that hormonal change, even when the hormonal change clearly influenced physiological characteristics that everyone associates with gender, did not change either the gender or the sexual orientation.

If one removes gender and simply discusses sexual orientation, hormones still are not the easy answer that might be expected. In the 1970 paper, he stated,

> Clinically, it is true that some homosexuals have a history of chromosomal error, sexual birth deformity, undescended testes, small penis, delayed puberty, gonadal insufficiency, poorly developed or contradictory sexual dimorphism of body build, gynecomastia (in boys) and hirsutism (in girls). But the frequency of any one of these disorders among homosexuals is so sporadic as to not create special hormonal research vigilance at the present time. Conversely, the incidence of homosexuality in the clinical population of each one of the listed disorders is so sporadic that one cannot seriously entertain the hypotheses of a primary hormonal cause-effect link between the physical symptoms, on the one hand, and homosexuality on the other. If there is any link, it is more likely to be secondary. There is far and away more homosexuality among organists, hairdressers, actors, interior decorators, or antique dealers than among patients with endocrine diagnosis! (1970g, 434)

Money listed the physiological effects symptomatic of endocrine disorders and noted that there is no correlation of sexual disorders – or, to put it in a less pejorative way, sexual diversities. This failure of correlation works both ways: those who have an endocrine disorder are not more likely to be homosexuals, and those who are homosexual are not more likely to have an endocrine disorder. And he ended with a characteristic bit of Money wit, something that all too often could be seen as flippant or even, as in this case, reasserting stereotypes.

Money enjoyed polishing off explanations that he found wanting, but he constantly looked for the possibility of explanations. In the 1961 article, he found consistencies in both family structure and morphology in the effeminate boys:

> In our study, the most consistently recurring findings were the infrequency of fatherly domination in the household, the lack of preference by the child for the father as the favorite parent, the relatively fragile body build of some of the boys, and as previously stated, the frequency with which the mother viewed the son's behavior in a more serious light than did the father. However, no simple point-for-point association of cause and effect of sissyishness or tomboyishness was revealed. (1961c, 3)

It might seem as though a search for a "simple point-for-point association of cause and effect" is overreaching, but, as he mentioned in many of his articles and books, such a "point-for-point" was often posited for homosexuality but also for many of the disorders he examined. While none of these possibilities, from hormones to parenting, satisfied any scientific standards for causality, causality remained the goal. Perhaps this was the reason Money returned again and again to intersex. The possibility of a person who is not wholly male or female, a possibility that seems to deny the simplicity of homosexuality or heterosexuality, might provide an insight into the key to sexual orientation.

Money began *Gay, Straight, and In-Between* with a chapter on "Prenatal Hormones and Brain Dimorphism." His primary concern was the various animal studies of dimorphism. There are many elements to such studies, but they tend to follow one basic model: something is modified on an animal of one sex in order to move it towards the behaviour of the other sex. Almost invariably, the success of this change is judged according to sexual behaviour. A male animal becomes female if it acts receptive to mounting, and a female becomes male if it tries to mount. Regardless of the accuracy of this assessment among animals, it seems obviously simplistic if applied to human gender identity. Gender for humans is much more complex than mounting or not mounting. It seems to me self-evident that we have many important markers of gender that are equally evident in persons of different sexual orientation. One might question why Money would open his book with animal studies.

His opening chapter might come close to the arguments of his opponents, such as Milton Diamond, who seem to believe that biology is destiny: by their chromosomes you shall know them. This view, that the sexual subject and gender subject are more accurately defined by the scientific observer than by the subject him- or herself, continues throughout many supposedly scientific studies. A good example of this attitude is Diamond's view of the plethysmograph, which judges sexual orientation by erectile response:

> While pupillary and genital responses are probably a valid reflection of erotic interest, the responses might also be to novelty, shock, or something else. Nevertheless, practiced test givers can most usually differentiate the responses of erotic interest from other causes. Many studies have documented the heuristic and practical value of such measures. (1998a, 63-64)

This scientific faith in the "practiced test givers" seems less than scientific. Many other studies have posited that the best way of defining homosexual is to just ask the subject, as suggested by the demographic research at UCLA's Williams Institute about those who identify as LGBT. Strangely enough, the testimony of a human voice is better than that of a penis.

Money, like earlier students of sexuality such as Havelock Ellis and Kinsey, was much more willing to accept the subject's autobiography. Still, autobiographies are of little value in the search for an explanation that at least suggests causality. Money made statements such as the following about an intersex woman:

> In the human species, the site where hormonally responsive brain cells are prenatally masculinized so as to induce a predisposition toward subsequent bisexuality or homosexuality has not yet been demonstrated. One must infer, on the basis of studies of laboratory animals, that the site of masculinization is not in the neocortex of the brain, but in the old brain, the paleocortex, also known as the limbic system, which is intimately connected with the hypothalamus. (1988a, 37)

Even with the advanced GPS systems of twenty-first-century science, we remain unable to find that "site of masculinization." For a fascinating, entertaining and very critical examination of this search for brain dimorphism, look at Cordelia Fine's *Delusions of Gender* (2010). Fine (2010, xxvii) says, "when we follow the trail of contemporary science we discover a surprising number of gaps, assumptions, inconsistencies, poor methodologies, and leaps of faith – as well as more than one echo of the insalubrious past." In a freewheeling interview with *Cosmopolitan* magazine, Money summed up his view:

> Before we're born, while we're under the control of sex hormones, certain things happen to our brains. If a girl baby is forming and nature adds some male hormone – testosterone – then what would have been an exclusively feminized brain is going to have a masculinized predisposition, and the chance of the girl's being bisexual or lesbian is greatly increased. With a male, the exact opposite is true. If *not enough* testosterone is added, the chance of bisexuality or homosexuality is increased. Of course the social experiences of the child may override those predispositions. But remember, while this is a very good hypothesis, it can be demonstrated one hundred percent only in animals. (1990b, 111, emphasis in original)

While Money's statement seems disposed to a prenatal hormonal explanation, it offers a very large loophole through "the social experiences of the child," which presumably could include anything. Even after that loophole, Money's final sentence still acknowledges the limitation of seeing humans as following the functions of animals.

I was quite delighted recently when one of the central purveyors of what might be called gay science, Simon LeVay, published a new book titled *Gay, Straight, and the Reason Why: The Science of Sexual Orientation* (2011). LeVay is likely to know Money's book of the similar title, but he makes no mention of it. LeVay's one reference to Money in *Gay, Straight, and the Reason Why* is the usual attack on his involvement in the Reimer case. But it amuses me that Money in 1988 offered only "and in-between," whereas in 2011 LeVay was ready to assert "the reason why." Well, not really. As LeVay (2011, 171) admits in his comments on Dean Hamer's famous study (1994) of the "gay gene," which looked for a genetic marker of homosexuality in males, "Unfortunately, Hamer's report has not been robustly confirmed." As every student of science knows, "robustly" is a standard modifier to assert that some study exceeds the possibilities of chance, but I still love the butch tone it produces. It seems to offer the hope of a macho recognition that all is as it must be. If sexual orientation had either a biological cause or a clear biological marker, then it would seem logical that some studies would have proven this by now. Instead, after almost three hundred pages of discussions of everything from chromosomes to finger length, LeVay (2011, 290) is able to conclude with only "biological factors may be closely involved." It is difficult to think of any aspect of human life to which this phrase would not apply.

Money had great faith in science, probably no less than that expressed here by LeVay. His belief in science was something wider, however, that included experience and society. Money always recognized the eagerness we all have to find a cause as well as the rarity of finding a cause that clearly produces an effect. As Greg Lehne recalled in conversation, if one of his postdocs suggested a simple cause for something they were studying, Money would say, "Do you have a crystal ball? Why don't I have a crystal ball?" Yet he had experience of both biological and social causes. For the former, increased knowledge of endocrinological deviations from the norm gave him the suggestion of strident work by biology against biology. Complete androgen insensitivity syndrome had shown him that it was possible for someone with completely male chromosomes who seemed completely female in almost every other way. His studies of abuse dwarfism showed that a significant physiological disorder might be not simply a response to physical conditions, such as lack of nourishment and physical abuse, but rather the direct physiological effects of psychological treatment.

In a 1964 review of a book that claimed "familial psychodynamics" to be the cause of homosexuality, Money showed how strongly he believed that science needed to recognize the larger picture:

> The authors' bias in favour of this inference shows up clearly in their handling of the heredity issue in their last chapter. Alas! They risk disgracing the professions of psychiatry and psychology in the eyes of other scientists when they show themselves still in the Nature-Nurture age of Herbert Spencer, unacquainted with the advances and concepts of modern genetics. Heredity is not a synonym for a blind kind of predestinarianism, a fatalism against which all therapeutic effort is futile. (1964a, 199)

Money's central argument about nature and nurture was this:

> It is counterproductive to characterize prenatal determinants of sexual orientation as biological, and postnatal determinants as not biological. The postnatal determinants that enter the brain through the

> senses by way of social communication and learning also are biological, for there is a biology of learning and remembering. That which is not biological is occult, mystical or, to coin a term, spookological. Homosexology, the science of orientation or status as homosexual or bisexual rather than heterosexual, is not a science of spooks. Nor is the science of heterosexology. (1988a, 50)

The holistic theory encapsulated in "a biology of learning and remembering" suggests the breadth of Money's approach to science. He refused a kind of geneticism that he described as fatalism, but he wanted a science that comprehended the biology of psychology and the psychology of biology. Money wanted scientific answers, and the most convincing answers about the body are those that can be linked to biology. Thus, he began with the animal studies and the hopes that they might offer. In the epilogue to his first chapter of *Gay, Straight, and In-Between,* Money wrote,

> Human sexological syndromes in the clinic represent experiments of nature that are the counterpart of animal sexological syndromes induced experimentally in the laboratory. Despite species differences and variations, data from these two sources are mutually compatible. (1988a, 49)

But as all these studies of homosexuality seem to miss sometimes, the human mind is very complex. The digestive systems of mice and humans are "compatible," but that does not explain why humans like French restaurants. What the narrator in Herman Melville's *Billy Budd* says of Claggart could be said of the human mind in general: "a nut not to be cracked by the tap of a lady's fan" (1967, 352).

From the animal studies, Money moved on to "Gender Coding." A good part of this section discusses transsexuals. This might seem a strange direction for a book ostensibly about homosexuality, but once again Money used his expertise in other areas of sexology to address sexual orientation. Thus, he stressed the importance of the subtleties of gender identity and gender role for the trans woman, who needs to find all aspects of her life to be understandable as gender coded. Money saw definitions of gender as invariably based on identification and complementation, or the process of saying "I am this thing because I am the same as X" and "I am this thing because I am not the same as Y." Money included sexual orientation in this process and therefore reinforced the idea that in our society a major part of gender coding is compulsory heterosexuality. Reminiscent of French feminist theorist Monique Wittig's claim (1992) that the lesbian is not a woman, Money suggested that the sexual other is a clear part of the idea of the sexual self. Money based this on a rather doctrinaire view of the world:

> In the theology of natural law, there is decreed an absolute standard of what is normal in sex, and that standard is procreation. If we were a species in which all the sexual acts of all males were exact replications of one another, and the same for females, then there would be an absolute standard by which to measure masculinity and femininity. That being not the case, we as a species must live with two standards of normality, the statistical and the ideological. (1988a, 76)

So Money descended from the doctrinaire, in which nature offers an absolute, with all of gender defined by sexual acts, through the scientific view of what should be, which is primarily statistical, to his own perception, which is the ideological. In this, as in so many things, Money's writing recognized the hegemony of the ideology of normality but, at the same time, worked subtly and overtly to press his own ideology.

Whenever there was an agenda in Money, it was usually towards sexual liberation. In *Gay, Straight, and In-Between,* he referred yet again to the importance of "sexual rehearsal play" in children, a key aspect of his lovemap theory. Money claimed that cultures that allowed overt sexual play among children had no sexual perversions and no homosexuality. But if Money was not some self-hating homophobe, why did he make such a claim? One of his constant examples was Aboriginal Australia. He claimed that his anthropological excursion to a mission in Arnhem Land was the source of his information. In one of his appearances as an expert witness, he testified, "I think one of the reasons why they had no sexual problems in a homosexuality [sic] was they have an ancient and very long tradition of being extremely openminded about sexual play in infancy" (1975a, 7-123). I have read his notes from the expedition, held in the Hocken Library at the University of Otago in New Zealand, and his observations seem to have been an act of will: in interviews he made sure he heard what he already knew to be the case. The likelihood is that there was less sexual rehearsal play than he claimed and that there was more homosexuality than anyone in that very Christian environment would tell him about. But Money needed evidence for the causality that would justify his pleas for the sexual liberation of children. It is tempting to look to biography and Money's own repressive fundamentalist Christian upbringing as a reason for this belief. He had examples from clinical experience of the negative, men whose unusual sexual needs correlated to childhood trauma. Regardless of Money's frequent questioning of causality, he often asserted a direct relationship between childhood sexual repression and adult sexual dysfunction. But he needed something positive to show the alternative, such as the absence of homosexuality in a world with children merrily exploring their sexuality.

If Money's claims about sexual rehearsal play and sex education for children seem somewhat homophobic, this could be simply polemics, as he tried to convince his perhaps homophobic audience of the need for freedom. Among the many discontinued projects that can be found in Money's papers is one from 1970 called "Significant Aspects of Erotica" (1970h). In this unpublished article, he asked,

> Is it better for a growing boy, walking in the park, to be greeted by a pleasant and friendly homosexual before or after having been advised of homosexuality and its implications for his future? Is the boy to wonder for the rest of his life what a homosexual encounter entails? Is he apt to try a homosexual experience in an effort to satisfy whatever curiosity he may have about sex with his own sex? Which is to be preferred: a physical encounter or a dialog leading to an understanding of the homosexual's way of life, a dialog inspired by the knowledgeable use of candid pornography depicting two men making love to each other? It is a good question, and worth pondering. (10)

This piece was perhaps rejected by a publisher or abandoned by Money himself; regardless, it represents the extreme of sexual openness that he would be attacked for so often. Invariably, throughout his career, Money believed that sexual well-being, regardless of sexual orientation, was based on as much early sexual knowledge as possible.

One of the central concerns in *Gay, Straight, and In-Between* that reflected sexual liberation but perhaps went beyond it is the statement that homosexuality is not a pathology but rather a parallel state to heterosexuality. The *Diagnostic and Statistical Manual of Mental Disorders* (DSM) had removed homosexuality as a general disorder in 1973, but "ego-dystonic homosexuality" – best interpreted as a homosexuality that causes the individual unwarranted stress – continued to be listed. Given the homophobia of society and the many homophobic laws in 1973, it is difficult to contemplate what might have been non-ego-dystonic homosexuality. Homosexuality was completely removed in 1986. Even this final excision, however, did not mean that everyone outside the North American gay community treated it as just part of the norm. Besides the anxiety of the world of the justice system, many psychologists continued to perceive it as a pathology. The ideas of earlier theorists such as Edmund Bergler (1956), who saw homosexuality as psychic masochism, were still prominent, and there were many contemporaries, such as prominent psychiatrist Charles Socarides (1995), who continued to state that homosexuality was a disease. There were many instances when Money felt the need to confront Socarides in print. The primary justification for what might be called homophobic psychoanalysis was that so many homosexuals were troubled by their sexual orientation. Money's response was that the problems were easily explained by social oppression, as in his analogy of left-handedness to point to something before pathologized and a focus of medical and psychological intervention but now generally accepted.

Money's opposition to homophobia was based on a number of factors. One was simply the error inherent in the views of people such as Socarides that the fact that the homosexual was distraught by his sexuality made it necessary for medicine to find a cure. In 1965, Money told an interviewer,

> I become quite incensed about this criterion of "wanting to be cured." I don't think medical people have any right to lay the blame on the patient as to whether he will be cured or not. It is their responsibility. Our basic job eventually, I will admit, is that we should know how to change a person's wants on an issue like this. Just to sit there and say we can cure you, if you want to be cured, I think is really malpractice. (1965b, 17)

The definition of the sickness cannot simply be that a person is convinced that he is sick. Medicine might see that the more appropriate cure would be to change the person's understanding rather than his condition.

In *Gay, Straight, and In-Between,* Money devoted a great deal of space to the paraphilias, what used to be called perversions. This once again might be seen to be questionable given his topic, but his central point was to exclude homosexuality from the paraphilias. He stated that homosexuality's "classification as a paraphilia is scientifically untenable, insofar as all of the forty-odd paraphilias may occur in association with homosexual, heterosexual, or bisexual mating" (1988a, 84). For Money, one of the central problems of the lovemap is the lust-love split, in which the paraphilia is considered by the subject to be an evil lust that must be separated from the love experienced in the subject's primary relationship. Money's position was that the lust-love split has nothing to do with sexual orientation. The lust-love dichotomy is created for some homosexuals by society's condemnation. Money enjoyed presenting this opinion in many courts as an expert witness, including two of the classic cases in the United States military, Leonard Matlovich in 1976 and Vernon Berg in 1977, discussed in Susan Gluck Mezey's *Queers in Court: Gay Rights Law and Public Policy* (2007). These cases led Money to comment in 1993: "It is conceivable in the national interest that the quack of the ugly duckling will be listened to, lest the United States military ban on homosexuals become the Achilles heel of national defense" (1993b, 23).

One particularly interesting example of Money as expert witness was his testimony in 1979 at the trial of the Toronto magazine the *Body Politic*. The publishers had been accused of obscenity for publishing a supportive portrait of the North American Man/Boy Love Association, an organization in the United States that works to abolish age of consent laws criminalizing adult sexual involvement with minors. The article's author, Gerald Hannon (1977-78, 11), suggested that men who felt erotic attraction to boys were "the heirs of Mr Atkinson, 'Leader in Boys' Work,' community workers who deserve our praise, our admiration and our support." C.J. Atkinson was founder of the Broadview Boys' Institute, later part of the YMCA. To call pedophiles "community workers" no doubt irritated many, but few would have been happy that Hannon seemed to suggest Atkinson was a pedophile. In his testimony on the case, Money went on at great length about masturbation, degeneracy, and even witchcraft to suggest that the *Body Politic* was the victim of societal prejudices: "There is always a temptation and a tendency to fall back on these antiquated worn out concepts simply because science hasn't entered into, filled the vacuum yet" (1979d, 252). No doubt, Money's agenda to remove legal restrictions on homosexuality was part of his general support for sexual liberation, but he also sought what might be called medical liberation from the assumption that doctors must treat "sicknesses" such as left-handedness, something that had proved to be neither damaging nor curable. As in the case of transsexuality, Money's approach was not to assume that someone troubled by gender or sexual deviance should have the deviance controlled, to reconfigure the patient to be "normal," but to recognize that the deviance needed a different response. In the case of transsexuals, this meant gender reassignment. In the case of homosexuals, this meant medical and legal acceptance. He said, in 1977,

> with sexuality, it is malpractice to enforce treatment on homosexuals and bisexuals ostensibly to make them heterosexual. If such treatment were enforced by law, the suffering of stigmatized people on the waiting list would be enormous, for the ratio of therapists to patients would be forever inadequate – to say nothing of the failure of outcome. (1977b, 232)

When the defence at the *Body Politic* trial asked about "a normal homosexual," Money offered what might be called his usual response: "the two correct definitions of normal ... at the midpoint of the normal curve of distribution ... [and] the ideological definition and ideologically society, or a family, or a group of people, have the norm of that which they themselves and their membership to behave by [sic]" (1979d, 252). He said of the latter, "that way of thinking about homosexuality does not tell us anything about either the somatic health or the mental health of the homosexual who is just as likely to be healthy as the heterosexual person is, in that sense" (253).

The problem of "normal" is an ongoing one. In one of his comments on intersex, Kenneth Zucker (1999, 1) was very careful to avoid the view of "abnormal" as "a monster," and yet he asserted that intersex is abnormal and that "it is hard to argue that such conditions are simply variants from the norm." But, in simple fact, all that "abnormal" represents is variants from the norm. It is up to society to decide what variants are so extreme as to inspire either medical or judicial action. Thus, while the article in the *Body Politic* represented abnormal behaviour, Mr. Atkinson's speech on "The Boy Problem" to the very establishment Canadian Club in 1909 was presumably the quintessence of normal. There he referred to "a red-headed, freckled boy, not much in appearance but who had a twinkling little eye which suggested much" (1909, 54). Suggested much what? I make this tangent to support the importance of Money's combination of the tenor of objective science and the assertion that homosexuality could be an acceptable variant from the norm. The transsexual required medical assistance to become what the patient considered to be normal. The homosexual required society to accept the patient as normal, to accept that the person should not be a patient at all. Money believed that there had been inevitable progress as society came to be less frightened by homosexuality but that the law, by continuing to treat homosexuality as a crime, was creating a time lag for the inevitable. In an interview with the American Psychological Association, he asserted, "It really does turn the clock back when in a quiet sort of way society was finding its way out of the morass it has been in since the era of the Inquisition on this topic" (1976c, 3).

Money's attempt to be dispassionate and scientific caused some confusion, especially when he turned to taxonomy. Thus, in *Gay, Straight, and In-Between,* he provided a chart of "gender crosscoding" (1988a, 85), which he divided into "continuous" and "episodic." In "continuous," "total" is "transsexualism," "partial unlimited" is "gynemimesis" (appearing as a woman), and "partial limited" is "homophilia." In "episodic," "total" is "fetishistic transvestism," "partial unlimited" is "nonfetishistic transvestism," and "partial limited" is "bisexualism." The chart itself represents a belief in taxonomy that overreaches the possibility. It seems akin to the anecdote about the zoologist Cuvier noted by mycologic historian Donald Rogers (1958, 332) in "The Philosophy of Taxonomy": "One day the devil appeared in his laboratory, complete with horns, hoof, and tail, and threatened to eat him up. Cuvier said, 'Quit bothering me. You're obviously strictly herbivorous. Go on outside and eat grass. You can't eat me.'" The question that should be posed to any taxonomy is the relevance of the claimed distinctions. Besides the problem of just figuring out Money's chart, many would question his continuum because they doubt that any kind of cross-dressing is intrinsically related to homosexuality. In his 1963 article, Money had seen some degree of cross-gender identity in even the most male homosexual: "a more masculine homosexual's feminism of imagery may be limited to the erotic arousal-power of a body with the same sexual morphology as his own" (1963c, 34). Once again, however, Money's assertion is about the impossibility of separating gender and sexuality. To see gender as simply one issue and sexuality as simply another is to deny our apparently ubiquitous belief that sexuality is understood through gender. No one knew the extraordinary range of paraphilias better than Money, and thus he would have been well aware of how far removed desire can be from a simple view of sexual orientation, but that desire still must be understood through gender. To begin with, the most extraordinary paraphilias are almost always exhibited by males.

In rather typically pugilistic style, Money attributed any disagreement with his analysis to "gay polemicists." He attacked them for "censoring basic homosexological science" (1988a, 153). But, as so often in Money's work, some basic observations and a rather freewheeling application of common sense took matters too far. Many have attacked any association between gender crosscoding and homosexuality. To them, variant sexual orientation does not suggest any variance in gender identity or gender role (Sánchez et al. 2009). Thus, including sexual orientation in the assessment of gender identity in intersex is just homophobic. But to return to Monique Wittig, one need not be an extremist of either the gay or straight variety to see that our ideas of gender are, as Wittig claimed, largely based on heterosexual assumptions – Cordelia Fine's "insalubrious past"(2010). Gender identity and sexual orientation are not the same, but neither are they separate. Yet Money accepted some degree of separation when he commented on the ease of identifying homosexuality according not to biological sex but to gender identity. In 1963, he suggested that, while the term *homosexual* would generally be thought of as relatively simple, it actually had multiple components: "In its ordinary usage, homosexuality implies that a couple are chromosomally, gonadally, hormonally, and morphologically homosexual and are assigned to the same sex" (1963c, 30). But in many cases, many of them invisible to the public at large, some of those components are not homosexual. Homosexuality must be based on only gender identity and gender role. Thus, a trans man with a female partner is clearly heterosexual, and a trans man with a male partner is clearly homosexual. The only quibble would be with someone who does not have a secure gender identity.

But regardless of Money's constant assertions that gender identity is larger than sex, he also constantly asserted that sex is central to gender identity. While his concept of gender identity is a major part of the reason why we now believe sex is not a property proven by simple observation, that does not mean that gender identity is a somewhat arbitrary matter shaped primarily by the needs of the culture. Instead, Money viewed gender identity as a base human attribute that can eventually be understood through careful scientific analysis. Yet there are various examples in Money's work not of careful science but of a leap of faith based on his analysis. As an expert witness in the case of L.M. Smith, fired from the United States state department for homosexuality, Money stated, "My evaluation of him, I may say in passing, he is not a very convincing homosexual. He is really a bisexual" (1975a, 7-25). This was not because of the subject's identity claims or the subject's sexual history but rather Money's own observations of his gender presentation: "I told him my evaluation of him was he was one of the many people who falls into the trap in our society of calling himself a homosexual when his mental potentiality is to be bisexual and I thought as time went on, he could explore and find out more about this other dimension in himself" (7-66).

I realize it might seem as though I have slipped from discussing sex to discussing sexuality, but this comment is very much about gender presentation.

Money seems to have thought that "a very convincing homosexual" would have a certain type of gender presentation. Someone who does not have this presentation is more likely to be bisexual, to have a sexual orientation that is not as clearly defined. This view of bisexuality as manifested in gender contrasts with bisexual experience, which, as Kinsey showed, is quite wide-ranging. Those most convinced of the biological truth of homosexual and heterosexual orientation tend to be least convinced of bisexuality. Diamond (1998a, 72) states, "When it exists, while it is much less common than heterosexuality or homosexuality, it is nevertheless, no doubt, the result of the same biological biases which determine either distinct androphilic or gynephilic arousal." Those such as Diamond who see sexual orientation as a biological fact are uncomfortable with an idea that bisexual potential might be part of many more people than those few who identify as bisexual. Those such as Money, who see a true homosexual as someone who is also deviant from the gender norm, see the bisexual as something like a homosexual who is not deviant from the gender norm. Then there are those such as Wittig (1992) who are able to see sexual orientation as an almost completely political choice. One need not wholly agree with Wittig to see that sexual desire could be ultimately much more dependent on complex intersections of behaviour, in which the human ability to choose figures very strongly.

Many of Money's concepts do seem commonsensical, such as the lovemap. One aspect of the lovemap that was given an unusual applicability in the context of Money's treatment of homosexuality is the trajectory of limerence. He suggested that "the natural history of a love affair is that the duration of the leaping-flame stage is around two years, after which the molten glow takes over. For the reproduction of the species, this is time enough for the pairbonding of the love affair to have progressed to sexual intercourse, pregnancy, and delivery of the baby" (1988a, 151). He noted that this was the same for childless couples and for gay couples. He presented the latter primarily to oppose the "shibboleth of promiscuity" (151), but is it accurate? Money seemed to claim that the two-year cycle represents just a general human condition. Perhaps a more apt suggestion would be that the long existence of the nuclear family has created a pattern that permeates the whole of society. According to Money, it is not just compulsory heterosexuality but also compulsory reproduction that shapes our lovemaps. In my teaching, I find it very difficult to convince undergraduates of the value of Wittig's view of the lesbian as somehow not a woman. Money's claim about reproduction is similar. There is an obvious biological truth that human life cannot be separated from reproduction, but neither can the metaphors of life. First comes love, then comes marriage, then comes the baby in the baby carriage. Even Money's "gay polemicist" will find it difficult to escape the way the pressures of such a constant trajectory influence our lives.

While Money's overall view of homosexuality was that of a sexual liberationist, the parts of the whole were much more diffuse. Back in 1963, he had referred to intersex individuals who are genetic males who "cannot be masculinized and so are assigned as girls. Then they grow up very well with the gender role and identity of girls, and are, if you will permit, experimental homosexuals of a sort, on the criterion of either chromosomal or gonadal sex" (1963c, 31). He would have been unlikely to use the term twenty-five years later, but the idea of the "experimental homosexual" suggests the temptation to seek an analysis beyond the claims of self-identified gay men and lesbians. While the term *experimental* might reinforce the common view of Money as the mad scientist seeking humans for studies of sexuality and gender, it also suggests the contradiction between life as it is lived, the woman who is born XY but has complete androgen insensitivity, and the assumptions of objective science, in which the chromosomes define what the human is. Very early in his career he recognized that life cannot be understood simply.

In "Factors in the Genesis of Homosexuality," Money presented a careful assessment of the possible process of becoming homosexual:

> In the genesis of homosexuality, it is an open question which is the stage in development from zygote to adult when the die is cast in favor of subsequent psychosexual differentiation along homosexual lines, whatever the degree of severity. There may be no one specific stage, but a variance in timing, upon which might depend the ultimate severity of the condition – whether chronic and pervasive (essential homosexuality), alternating and bisexual, or transient and circumstantial (opportunistic homosexuality). (1963c, 37)

I can quibble with the pejorative tone of some of the words, but this remains the best encapsulation I have read of the scientific approach to homosexuality. Regardless of the variety of claims about how and when it begins, same sex orientation is a given. One might not refer to "degree of severity," but "degree" should be a part of the understanding as there seems to be no limit to the range of homosexual and heterosexual experience among men and women who could be called "homosexual." And at least in this instance, there is no comment on cause.

Regardless of his claims to dismiss causality, Money was as drawn to speculation as any of us, as shown elsewhere in that same article where he provided an emphasis that angles towards a potential:

> Virtually nothing is known of the functioning of the special senses in relation to the formation of gender identity. From studies of imprinting in animals, it is known that some behavioral traits and activities are set in motion by the release of a neuromuscular mechanism that is phylogenetically preset to be released by a suitable perceptual stimulus ... It is conceivable that some analogue of imprinting takes place in the establishment of human gender role and identity. If so, then the possibility exists that boys and girls may be differentially equipped phylogenetically to distinguish male and female stimuli and to respond to them as objects of identification. (1963c, 37)

Money often asserted the validity of understanding some human behaviour as imprinting. His word *conceivable* suggests an extreme hesitation, one for which conception is not perhaps the best representation, but the idea of imprinting is a useful one for sexual orientation. If the many vagaries of human agency are added, this seems to be a possible explanation of what might be called preset homosexuality, a preset that in some cases might never receive sufficient imprinting and in others might be imprinted by the slightest encounter. And regardless of the variance in response, in spite of LeVay's claims, there is still no convincing explanation of why that "preset" exists.

Yet, while this might be the case, it could be that this phylogenetic response is not gender-specific. It is possible that this unidentifiable mechanism, that cause so yearned for by scholars such as LeVay, is the same for all humans, male, female, or intersex:

> A still more likely possibility is that boys and girls are phylogenetically the same in their preparedness to establish a gender role and identity, being equipped only to impersonate an identity with a stimulus object. The decision whether this object will be male or female is not phylogenetic at all, but a product of ontogenetic opportunities and rewards. (1963c, 38)

While "ontogenetic opportunities and rewards" might seem to suggest scientific specificity, they might be translated as "Someone is homosexual because he is homosexual." We all have ontogenetic opportunities and rewards, some of which are quite similar for all humans, but different humans respond differently. Money's final words in the 1963 article read thus:

> One arrives at the conclusion that the final common pathway for the establishment of a person's gender identity and, hence, his erotic arousal pattern, whatever the secondary and antecedent determinants, is in the brain. There it is established as a neurocognitional function. The process takes place primarily after birth, and the basic fundamentals are completed before puberty. The principles whereby all this happens await elucidation. (1963c, 41)

We continue to await elucidation. The search for sexual orientation as a "neurocognitional function" might give rather too much credit to the brain over the amorphous philosophical construct known as the mind. In Shakespeare's words, "Love looks not with the eyes but the mind." Yes, but how does the mind look?

CHAPTER SIX

The Edge of the Alphabet

Neologisms

The author of *The Edge of the Alphabet* (1962), Janet Frame, was no doubt Money's greatest friend over his life. There were many reasons for the depth of that friendship, some explored in Michael King's biography of Frame, *Wrestling with the Angel* (2000b). In the beginning, Frame seems to have been a naïve young woman fascinated with the apparently worldly Money, a romantic crush. Money's reciprocal fascination was as a nascent psychologist. As a young professional overwhelmed with possibilities, he found Frame to be a brilliant artist who was deeply troubled in ways that he thought he might be able to help. As their lives developed, they shared a fascination with New Zealand combined with a revulsion at what they saw as its parochialism. But perhaps there was another bond, not noted by many. As the title of her 1962 novel suggests, Frame was compelled by and tortured by language, by what it could and could not do. Perhaps surprisingly for a psychologist, Money was the same.

This chapter might seem out of place in a book that devotes so much attention to medical treatment and psychological understanding of gender and sexuality. Yet, as I state in the introduction, this is very much a text-based study. I explore Money's theories as they appear in his various publications. While the concern ultimately is always human gender and sexuality, the reality of life as it is lived, I seldom stray from the written representations. As so many theorists have stated, in a variety of ways, language is ultimately self-referential. We might see the referent behind the sign, but it can never be there in a simple one-to-one correspondence. The sign is only the sign. This is the problem constantly addressed by legal theory, as lawmakers attempt to use language that can define human behaviour, can say what is allowed and what is not, can say what is and what is not. Thus, Money yearned for something that went beyond the mot juste, that could both define the behaviour and suggest the medical possibilities of understanding and even of treatment.

Some neologisms assert a rapid power that seems like a universal forest fire: *Internet* is a classic example. Others are created and then die a quick death – like that new word that you read last week and can no longer remember. The process of creation is the same: the creator decides that something has not been named, or at least not sufficiently or properly named, and so decides to do the naming. This is the Adamic delusion: the belief that the world is unnamed and God has given you the job. The hubris is palpable. Do you know so much that you can be sure no one else has already named it? How do you know that you are the best person for the task? How can you assume that your name will stick as opposed to other possibilities? And perhaps most important, is there a need for a specific name for this specific thing?

Money's most successful neologism was the simplest and the first and, in a sense, the least obvious. The title of the present book is both a praise of Money and a slight joke at his expense. His consistent claims that he originated the ubiquitous term *gender* at times seemed like an obsession. The extent of his assertions is clear from his 1984 lecture, "The Conceptual Neutering of Gender and the Criminalization of Sex." He stated, "Today the term *gender* in idiomatic English and in translation, applies to human beings in such expressions as gender role, gender identity, and gender gap. Thirty years ago, however, these expressions did not exist in any language" (1985a; 1986d, 591). He quoted his original use to explain the way intersex persons operated in the world within a gender regardless of the contradistinction of genitalia or chromosomes:

> This definition reads thus: "The term gender role is used to signify all those things that a person says or does to disclose himself or herself as having the status of boy or man, girl or woman, respectively. It includes, but is not restricted to, sexuality in the sense of eroticism" (Money, 1955, p. 254). (1985a; 1986d, 593)

Money then traced a series of modifications, such as Evelyn Hooker suggesting the term *gender identity* (1965) in one of her many studies questioning the assumption that homosexuality was a mental disorder, and then Robert Stoller dividing *sex* from *gender* in his book *Sex and Gender: On the Development of Masculinity and Femininity* (1968). Money perceived Stoller's division to be one of the primary sources of the loss of the biological and the sexual from his concept of gender role: "The neutering of gender is part of a tidal wave of antisexualism that spreads its octopus tentacles into the politics of what could become a new and dictatorial era of antisexualism" (1985a; 1986d, 598).

The power of Money's concept is generally acknowledged by sexologists. A good example is the view of Vern Bullough, the historian who was probably the best synthesizer of sexological ideas in the late twentieth century. He credited Money's definition of gender with offering a "totally new world view" (2003, 231). Bullough stated, "In a sense, by adopting a new term to describe a variety of phenomena, Money opened a whole new field of research" (232). Bullough then considered the ways the word came to mean a variety of things, some in direct contradiction of Money's original definition:

> This substitution of the word gender for biological sex should serve as a sort of moral story indicating that it is not always possible to predict what others will do with your specific terminology once it is disseminated – something about which we should all be aware. What we need now is another term to express what Money so ably explained nearly half a century ago. (232)

Money was frustrated by these changes but also by what he saw as the failure outside sexology to credit him with this very early achievement. Money's 1955 paper and its definition of gender role were the origins of his whole career. All other theories and honours built on that beginning. The level of anger he felt is suggested by the following in *Gendermaps* (1995):

> In its second edition, the *OED*'s first illustration of *gender* as a "euphemism for the sex of a human being" is taken from the book, *Sex in Society* by the British writer Alex Comfort (1963): "The gender role learned by the age of two years is for most individuals irreversible, even if it runs counter to the physical sex of the person."
>
> Not only is this usage wrongly called a euphemism, but also its year, 1963, is eight years too late for the first appearance of the term

> *gender role.* Alex Comfort borrowed the term from the works of John Money who coined and defined it, as will be recognized "when the entries for *gender* and related terms come up for revision" in the third edition of the *OED* to be "published sometime early next century," according to a personal communication to William P. Wang (March 12, 1991), from the senior assistant science editor of the *OED*. Its first appearance in print was in the paper (Money, 1955) on "Hermaphroditism, gender and precocity in hyperadrenocorticism," published in the subsequently discontinued *Bulletin of the Johns Hopkins Hospital*. In this paper the word *gender* made its first appearance in English as a human attribute, but it was not simply a synonym for sex. With specific reference to the genital birth defect of hermaphroditism, it signified the overall degree of masculinity and/or femininity that is privately experienced and publicly manifested in infancy, childhood, and adulthood, and that usually though not invariably correlates with the anatomy of the organs of procreation. (1995, 19)

It is not overstating the case to say that this lengthy account bristles. While the detail is no doubt accurate, the overall effect has not been as Money expected. As of 2014, the *OED* has not significantly changed the entry. Comfort is still there for gender role, but he is preceded by this: "1957 *Arch. Neurol. & Psychiat.* 77 333 *(heading)* Imprinting and establishment of gender role." Money is given no credit, although he is listed as the first to use the phrase *gender-neutral,* in 1963.

It seems as though Money is still being short-changed. However, under *gender*, the *OED* has another entry: "1945 *Amer. Jrnl. Psychol.* 58 228 In the grade-school years, too, gender (which is the socialized obverse of sex) is a fixed line of demarkation, the qualifying terms being 'feminine' and 'masculine.'" In other words, it seems that the version of *gender* that came to dominate our language actually predated Money's usage by at least ten years. He seems to have tapped into an understanding already existent in American psychology, although he no doubt narrowed and refined it. Over time, due in no small part to Money's own work and to his personality, his 1955 usage became an originating force in sexology, the tributary from which many streams developed. In the end, however, *gender* seems to have returned to where it began, as it was used in 1945 to represent "the socialized obverse of sex."

This brief history of the term says a lot about Money and his role. He might have taken credit for a term already in use, and clearly offered no doubt as to his position as originator, but he also made better use of the term, turning it into something that, as Bullough noted, was dearly needed. As the term left his control, and an etymological source such as the *OED* did not credit him with first usage, he became incensed and felt that once more he was not given his due as the founder of a major idea. The need for such credit was evident in an interview with the gay newspaper *Christopher Street* in 1980, in a comment on *ambisexual:*

> In their study, Masters and Johnson credit their adoption of this term to Mark Schwartz, who had completed post-doctoral training with me at Johns Hopkins. There, in the early fifties, I may have been the first or among the first to use the term, applying it to individuals with sex-organ anomalies such as hermaphroditism. Masters and Johnson use it in a new context to mean something about behavior. I don't use the word in *Love and Lovesickness* because it wasn't relevant in my vocabulary. Therein, what is its virtue? What does *ambi*-sexual say that *bi*sexual does not? Maybe 'disexual' or 'duosexual' will appear next. (1980a, 26)

Money was certainly not the first to use the term *ambisexual.* The first use cited by the *OED* was 1938. It is interesting that Money should have sought credit and then dismissed the word, a word that he often used, and then suggested it is just as silly as other possible neologisms. His first interest in language seems to have been his own control of its dispersal. Money's approach to language was very similar to his approach to sexological theory – eager and aggressive creation and control.

Money's concern for the importance of language was visible from the beginning of his career. In 1949, he wrote a paper entitled "A Specific Problem in Methodology: An Example" (1949a). His belief in his ability to change basic terms in social science appeared in a footnote to his use of *units-human:* "This neologism is used to avoid clumsy circumlocutions, and to provide a term which carries a connotation of depersonalization, in order that it may more clearly be differentiated from the concept of role. Hence the stress on unit: unit-human and not human unit" (1949a, 3). On the other hand, he was quite ready to be conservative in his approach to language when it suited him, as in an otherwise positive 1964 review of a book on homosexuality:

> Quotation marks are overused on everyday words that do not need them, in what Webster calls the equivalence of affixing *so-called* to a term. This error is alarmingly widespread in psychiatry, psychology

> and social sciences. It detracts from the dignity of writing and creates the impression that the author doesn't know, or isn't sure of what he's talking about. (1964a, 200)

Of course, the use of "scare quotes" means "so-called," but it also means that the term should be seen as having something more than the obvious or the expected meaning. It suggests that the reader should not fall into the trap of assuming that the sign and the referent are in an easy comfortable relationship. While it no doubt does not have the "dignity" that Money maintained in both his writing and his professional demeanour, it represents the motility of words and phrases. Money himself tried to use neologisms to fix words, to use them to assert his own perceptions of what science should be stating.

In an article in 1979, he used etymology to attack an American Supreme Court ruling that pornography was defined as "appeal to the prurient interest." Money observed:

> Etymologically, prurient means itching, and also to have a longing for something. In contemporary usage, when the connotation of the longing is sexual, it is immoral or indecent. Webster's dictionary defines prurient not as amative, romantic, or erotic, but as lustful, lewd, or lascivious. Here the law is blatantly tautological, for it connotes as indecent that which has prurient appeal, and prurience has the connotation of indecent appeal. (1979c, 28)

Money's argument worked as an observation on dictionary meaning, although offering little or no reason to oppose the law itself. He went on to suggest that this error leads to discrimination against homosexuals, but it is difficult to see why *prurient* makes the law any more or less discriminatory. A tautology does no more than restate what has already been asserted. It does not change the effect of the original statement, except, perhaps, to emphasize it. But just as Money saw words as sources of prejudice, he thought they also could be sources of liberation. In a 1979 review of Leslie Fiedler's book *Freaks: Myths and Images of the Secret Self,* Money maintained that Fiedler's desire to destigmatize *freaks* was not the answer. Instead, he suggested that the term be replaced with one of his favourite devices, a borrowing from a classical language, in this case the Greek *Katharma*:

> A person whose social stigmata need to be cleansed by society so that he or she may become a rightful member of the human race.

> Here is the new word we need. The degree of acceptance accorded to the unique individual, the Katharma, may be a measure of our maturity as a civilization. (1979a, 109)

Money's concern for etymology was consistent but not consistently rigorous. As in his own research, Money expected a higher standard of precision from others than he maintained himself. Thus, in his book *Gay, Straight, and In-Between,* he stated,

> Like the term gender, the term gay had entered into the universal English vocabulary by the 1960s. In the preceding decade, it had belonged in the vocabulary chiefly of the homosexual community. Its origin is obscure. It may have been adopted from the name of the historical era, the Gay Nineties, which were known not only for their frivolity in entertainment, but also for their partial sexual liberation from the dour Victorianism of the times. (1988a, 152)

The *OED* defines *gay* in the second meaning as "addicted to social pleasures and dissipations." The earliest use given is 1597, but the most obvious is 1754 from *Adventurer:* "The old gentleman, whose character I cannot better express than in the fashionable phrase which has been contrived to palliate false principles and dissolute manners, had been a gay man, and was well acquainted with the town." The use of *gay* for a female prostitute began in 1795. The first specific written use of *gay* to mean "homosexual" was Gertrude Stein in 1922, but judging from the language she was using a term that was already in general use. Some of this detail does not appear in the 1971 *OED,* the edition that would have been available to Money, but there is certainly sufficient information there, and in the other major dictionaries of the day, to take the origin well beyond "the Gay Nineties."

In some ways, *lovemap* represents Money's neologisms at both their best and their worst. It seems on the surface a bit silly, but it also concisely and clearly represents the concept. This was always very important to him. In a conversation with me, the psychologist Tom Mazur recalled a two-hour meeting when he was a postdoctoral fellow to discuss something for which Money wanted a new word. He said there were ten other fellows and colleagues in the room, and he thought, "How much is this costing?" Similarly, Money constantly used the term *sexosophy,* suggesting it had been adopted as a word for those attitudes not sufficiently scientific to be called sexology. In his plenary to the twenty-fifth anniversary of the National Institute of Child Health and

Human Development, he presented his development of *sexosophy* as one of his central accomplishments (1987b). While there have been a few uses of it, the only professional who adopted it was Money. One scholar suggested to me that others used it before Money, but I have been unable to find any reference. It seems possible that, as in so many instances, Money used a word so often, in a manner to suggest that it is common usage in the field, that he convinced others that it existed as an accepted professional term, not just another Moneyism. In the long term, however, again like other such words, *sexosophy* seems to have had no life beyond Money.

His primary achievements with words have often been rather to take those coined by others and make them part of psychological and medical vocabulary, such as *limerent*. A more important example is *paraphilia*. Anke Ehrhardt (2007, 223) credits Money with convincing the change from "perversion" to "paraphilia" in DSM-III. This change undeniably created an important transformation in the treatment of sexual deviance. There were other examples, however, in which Money's logic was probably accurate but the language has not been adopted. For example, in an interview in 1987, he suggested avoiding the word *adolescent:*

> The thing we call adolescence is a cultural invention, says John Money, the only scientist who has had the research support from the National Institute of Child Health & Human Development (NICHD) for each year since the institute was founded 25 years ago. "Just like the new wave of community colleges ... adolescence was intended to keep teenagers dependent on their families and out of the job market." Money told *Adolescent Magazine* he prefers the adjective ephebic to adolescent because it only refers to a development period in a person's life, not to a cluster of cultural beliefs about that time. (1987a, 1)

A similar example, the word *eonism,* created by Havelock Ellis, is now not used, but in a 1968 unpublished article Money claimed its value on the basis of precision:

> One cannot look to the chromosome count for an explanation of people like the Chevalier d'Eon de Beaumont, a nobleman and diplomat of Louis XV, who dressed and lived as a woman for years, but at autopsy was proved to be anatomically a normal male. Transvesticism has been used synonymously with eonism as the name for

> this contrasex syndrome of passing as a member of the opposite sex. A new term, transsexualism, has also appeared to refer to those eonists who achieve a formal reassignment of sex following cosmetic surgery and hormonal treatment – well known through the publicity of the Jorgensen case. Eonism is the more comprehensive and accurate term, since transvesticism ... may occur sporadically and without the person permanently assuming the opposite sex-role; and transsexualism does not entail a transformation of sex phenotypically but also cosmetically, superficially and incompletely.

Money was asserting that the term that should be maintained is the one that is most successful at describing all aspects of the subject under discussion. As he noted, "transvesticism," as simply "cross-dressing," is specifically about clothing and need not imply any representation of gender. It would include, for example, a man wearing female underwear but dressing outwardly as a male. On the other hand, the transsexual goes far beyond simply cross-dressing. *Eonism* could be a term that incorporated all versions of "passing as a member of the opposite sex." On the one hand, this might seem to be creating a blanket term where none is required; on the other hand, there are a number of instances in the twenty-first century in which various aspects of "transgender" representations are grouped together and in which *transgender* is an uncomfortable collective word but the only one available.

There are many examples of Money's play with language, and some of them seem to be gestures of little value, such as "homosexology" and "heterosexology" in *Gay, Straight, and In-Between* (1988a). I would think that his own work showed that this bifurcation of sexology is neither accurate nor useful. In the same book, he went a step further: "That which is not biological is occult, mystical or, to coin a term, spookological" (1988a, 50). He presumably coined the term to suggest how absurd a non-biological view might be, but the effect, at least to me, is to suggest that, as the user of the term, he became someone who dismissed ideas without consideration, like those who so easily label others "fascist" or "communist." While the term exists to dismiss those who are not scientific, *spookological* suggests someone who is less than scientific himself. The central argument of Donald Rogers's article "The Philosophy of Taxonomy" is that taxonomy is a science. The article begins with a definition of taxonomy:

> Taxonomy is the science of classification, and includes not merely the identification of individual organisms, not merely their arrangement

> in species and genera and more inclusive groups, but the whole basis of comparative morphology and physiology, and the whole framework of phylogenetic hypothesis, on which classification is developed. Ultimately, taxonomy is one sort of synthesis of almost everything that is known about living things. (1958, 327)

Just as much of Money's work, such as the lovemap, was about patterns of organization and classification, so too was his search for new words. In his article "Paraphilias: Phenomenology and Classification" (1984b), he offered a table of thirty-three paraphilias, including nine that were identified as "new terms formed from Greek roots (in collaboration with Diskin Clay, Professor of Greek, Johns Hopkins University)" (166). Some are quite extreme, such as one for "penile exhibitionism": *pedeiktophilia.* Given that almost all exhibitionism involves displaying the penis, it is difficult to see what has been added here. Then again, one term, *apotemnophilia,* seems to have become the standard label for the obsession to amputate a limb.

As suggested by his concern for taxonomy, many of Money's innovations with words claimed to produce a more scientific analysis of problems but often seem more concerned with employing language that made his analyses appear more scientific. As a psychologist, Money had a scientific designation. With that designation, he had considerable authority when he was a spokesman for the Johns Hopkins clinic, but he usually considered himself a "sexologist," a category that in the twenty-first century is still yearning for scientific respectability. Its role in language is suggested by a quotation from *Allegory of Love* by C.S. Lewis in the *OED* entry: "The second factor is the medieval theory of marriage – what may be called, by a convenient modern barbarism, the 'sexology' of the medieval church." In other words, not only did the field need defining, but even the term needed to be moved from "a convenient modern barbarism" to become a word for a knowledge accepted as a basic part of the human sciences.

This quest for legitimacy could be the reason behind Money's use of a term such as *pair-bonding,* which took a compound usually employed in zoology in order to make human relationships appear more biologically organized. We tend to see animal sex as inevitable, which suited Money's argument about the human lovemap. Thus his explanation of *phylisms* in *Gendermaps:*

> The basic building blocks of segments or sequences of behavior belong to all human beings, male or female, not as products of social construction, but simply by reason of our evolutionary heritage as

> members of the human species. They have been variously named as instincts, reflexes, and (in ethology) innate recognition and releasing mechanisms or fixed action patterns, but they have lacked an overall generic name. To overcome that lack I settled on the term phylism (Money, 1983). This term is derived from the Greek, *phylon* (Latin *phylum*) and is related to phyletic and phylogeny. A phylism is defined as a unit or building block of our existence that belongs to us as individuals through our heritage as members of our species. (1995, 36)

All aspects of this quotation are exemplary of Money's neologisms. He established the lacunae, used himself as a source, and explored the etymological connections to create a term for human behaviour that connects to the evolutionary and taxonomic spirit of *phyletic*. He saw these phylisms as covering all aspects of human action, including "shitting, fucking, laughing" (37). He went on, "Other phylisms have Latinate names, like thermoregulation, salt regulation, and immunoregulation. Still others exist that have yet to be named, or that have been named only recently, for example, pairbonding and troopbonding" (37). Money certainly had a smooth approach to his innovations, as he used his concept of the phylism to justify his use of *pair-bonding* and vice versa. The hermeneutic circle in which the whole explains the part and the part explains the whole became here a system of justification as well.

In 1955, Money published an article entitled "Linguistic Resources and Psychodynamic Theory," in which he began with a dictionary reference: "Present-day psychological vocabulary includes new word forms, like aggressivity and compulsivity, which are not listed in the 1951 edition of *Webster's New International Dictionary*, nor in the *Oxford English Dictionary*" (1955d, 264; 1986d, 69). Money offered an argument for the necessity of these developments:

> Some people denounce these word forms as perverse technical jargon, debased duplications of more traditional forms such as aggressiveness and creativeness, instinctive and perceptive. In point of fact, they are not debasements for they conform to linguistic precedent. Nor, and this is one of the arguments of this study, are they duplications. (1955d, 264; 1986d, 69)

In an imaginative move typical of Money, one with a clear comprehension of the hermeneutic process, he suggested that the process of understanding in psychology is perhaps as Latin-based as the words:

> There is also no *thinkuality*, *dreamance*, *knowivity*, or *choosivity*. Perhaps this is why in contemporary psychology thinking, dreaming, knowing, and choosing are virtually disregarded as dynamic determinants. Motivation is patrician! It requires Latin origins as well as endings. (1955d, 266; 1986d, 70)

Clearly, Money argued here for the necessity of a certain kind of Anglo-Saxon, non-Latinate understanding. More important for my argument here, he recognizes the interdependence of language and theory or, still broader, language and understanding. Money's search for words might seem irrelevant or even a waste of time, and his suggestion that all of his neologisms would be adopted by the scientific community absurd hubris, but his assumption that words shape everything was essential to his personal successes and failures as well as the successes and failures throughout the study of gender and sexuality. The understandings implied by *transsexual*, by *trans man* or *transman*, are intrinsic to these communities and to the medical treatment of anyone who could in any sense be called transgendered.

Regardless of this early awareness that including the Anglo-Saxon might offer a greater variation in thought, Money himself usually turned to Latin and Greek for his neologisms, as in the case of the paraphilias. This is true of perhaps my favourite Money neologism, the *paleodigm*. He introduced it in an article in 1989, in which the title suggests his assumptions as to its importance: "Paleodigms and Paleodigmatics: A New Theoretical Construct Applicable to Munchausen's Syndrome by Proxy, Child-Abuse Dwarfism, Paraphilia, Anorexia Nervosa, and Other Syndromes." *Paleodigm* seems to offer unlimited potential. He presented it three years later in his book *The Kaspar Hauser Syndrome of "Psychosocial Dwarfism": Deficient Statural, Intellectual, and Social Growth Induced by Child Abuse* (1992a) with all the pride of a new parent:

> The newly coined, newly defined term, paleodigm, came into being to fill a conceptual void and provide a name for the irrational absurdity of what it is that a mother is doing when she becomes a cruel tyrant who presides over the systematic sacrifice and possible death of her child. Paleodigm is defined and explained as follows (Money 1989, 15):
>
> > *Paleodigm:* an ancient example or model of a concept, explanation, instruction, idea or notion, preserved in the folk wisdom of

> mottos, maxims, proverbs, superstitions, incantations, rhymes, songs, fables, myths, parables, revered writings, sacred books, dramas, and visual emblems (from Greek, *paleo,* ancient + *deigma,* example, from *deiknyai,* to show). Paleodigmatics is the organized body of knowledge and theory of paleodigms. (164)

Paleodigm seems to be a perfect word for something such as child sacrifice, perhaps best exemplified in the story of Abraham and Isaac. One might consider this in the light of such well-known concepts as the Oedipus complex. Freud was well aware of the potential in mythic narrative, the suggestion that an ancient story might offer an understanding that psychology has not quite reached. To have a term that encapsulates the generality of that understanding, of the opportunity in centuries of story-transmission to convey a paradigm that continues in spite of limited theoretical analysis, would seem more than useful; as with Money's other neologisms, however, the word has not come into common use. As lexicographers often note, anyone can create a word, but the people will decide what is going to be used and thus what will enter the dictionary.

Although Money was attracted to complex formations that seemed sophisticated enough for medical conversations, he also recognized the need for simple language. Claude Migeon and Tom Mazur both noted his tendency to describe the womb as a "baby nest" when talking to children. Dr. Migeon suggested that "baby nest" was more than a cute phrase; it represented an unusually successful example of how pediatrics should work, especially in complex and difficult areas such as sexuality and endocrinology. On the other hand, Money was never afraid of words that others might find vulgar or even obscene, such as the phylism noted as "shitting." In *The Adam Principle,* he was able to embrace both *fuck* and Greek in a quite entertaining peroration:

> The honest Anglo-Saxon fuck, now living in disgrace, was once as wholesome as eat and drink, or walk and talk. If it could be rehabilitated, it would be just the right word, with exactly the correct meaning, from which to name the fuckological component of G-I/R [Gender-Identity/Role] – the component that classifies whether you do it like a man or a woman, homosexually or heterosexually, regardless of whether you were born with or without a birth defect of the organs with which to do it.
>
> For many years, no matter where I searched for the etymological origin of the once unspellable word f---, the answer was always the same: Unknown. In 1987 I found the answer (Money, 1988) [1988a].

> The Greek ancestor of fuck is *phuktos*, a gerund belonging to the verb *phutuein* (aorist passive, *ephuchthen*), which means to sow or plant in the ground, hence to beget and to impregnate. (1993a, 43)

Money's etymology is not supported by the *OED* or by other sources I have examined, which tend to emphasize Dutch or Scandinavian roots, but this says nothing about the exuberant spirit with which he pursued his study. In the article, Money then added excursions through German, Latin, Italian, and French before he concluded,

> Today, people would perhaps not find it too offensive to see the word differently spelled, in keeping with its Greek rather than its Latin origin. Then, with a word with which to name a new science of phucktology (or perhaps, phukteology) it would be possible to fill the void that presently exists and to set about creating that new science.
>
> The day might then arrive that one would open the dictionary and find meanings for an array of new terms belonging to that science: phuck, phucking, phukic, phuckically, phucktuous, phucktuousness, phucktuance, phucktual, phucktuality, phucktitive, phucktition, phucktitional, and phucktivity. Not only would the language of science be enriched, but so also would the depth of our understanding of our own experience of our own personal G-I/R as heterosexual, homosexual, or bisexual in its phucktuality. (1993a, 43)

This peroration is first playful, but that assertion in the last sentence shows the larger purpose: first to "enrich" "the language of science," but second to deepen "our understanding of our own experience" with the suggestion that active sexuality is central to our gender identity and the role that this gender presents to the world. While *phuck* might seem patently ridiculous to many, notice how often we use the term "the f-word," a euphemism that most seem to find acceptable but leaves no doubt that it stands for a word that most find unacceptable in polite discussion, particularly in its written form. This might support Money's notion that a new spelling, which made it no longer "the f-word," could make the word less off-putting for many. Note at the beginning the word *fuckological,* moving the already tenuous position of sexology into a field that presents an overt expression of the importance of the physical action of sexuality. At the same time he was suggesting the "f-word" should become a "ph-word," he was trying to make the world of science accept the original. *Fuckology* was one Money neologism that he presumably did not really expect to be adopted, but he used it often and at least once in the title of a conference paper, "Fuckology: The Science We Lack," which he presented to the American Association of Sex Educators, Counselors, and Therapists in 1996.

In 1978, Money published "Dirt and Dirty in Sexual Talk and Behavior," in *MIMS,* the *Monthly Index of Medical Specialties,* which is, as the name suggests, a trade journal "for Health Care Professionals only." I note this because it is yet another example of Money proselytizing wherever he found a venue that might broadcast his message more widely or to a new audience. In this case, he began with quite a lengthy examination of the etymology of *dirt.* He made a brief excursion through Nigerian usage before he reached *fuck* and his usual confrontation with its banning:

> The taboo on the word *fuck* has left generations of people whose native language is English without a publicly usable verb or noun that fits into everyday usage as colloquially as do eat, sleep, think, talk and dream. That deficiency is not fortuitous, for the very nature of a taboo is to proscribe an activity in which human beings might otherwise ordinarily engage. (1978b)

Money asked society to pay attention less to the continuation of Victorian prudery than to the forgotten undercurrent of sexual innovation:

> The Victorian era was one in which sexual taboo co-existed not only with the discovery of modern birth control but also with the new morality of sexual freedom thereby made possible. The beneficiaries of this new morality talked in their own vernacular – illustrated with its pictorial counterpart – that was more idiomatic and down to earth than the stilted terminology of medicine and the law. (1978b)

Money concluded, "Now is the time to clean up the dirt and to equate the dirty with the earthy" (1978b).

In the prologue to *Venuses Penuses,* in which Money described his development as a psychologist, he stated, "The argot of psychology obfuscates the meaning" (1986d, 14). He went on to assert his "resolve to avoid jargon talk" (14). The term *jargon* has a number of meanings, but the most common one today, the one intended by Money here, is "unfamiliar terms, or peculiar to a particular set of persons" (*OED*). It is difficult to reconcile this with his own practice: Money loved jargon and creating jargon. He seemed to have no idea in sexology for which he did not want to find a Latin or Greek word. His desire to affirm language that would make sexual knowledge accessible to anyone was met by his taxonomic obsessions, to make each aspect of gender, sex, or sexuality into an identifiable area of study. Still, Money's intention was never to use the unfamiliar or to communicate only with the particular. He wished to make the unfamiliar familiar and to extend all possible knowledge well beyond the particular set.

A representative example is a piece he wrote in 1982 titled "To *Quim* and to *Swive*: Linguistic and Coital Parity, Male and Female" (1982b). He began, "In neither the standard English vocabulary of literature and science, nor the vernacular vocabulary of uncensored speech, are there terms by which to distinguish what the woman does to the man, in the procreative act, from what the man does to the woman" (173). He came up with this answer:

> As a starter, I have conducted an extensive library search for two words by which to distinguish the female component from the male component of penovaginal interaction. The two words that sound the most idiomatic and least neologistic are *quim* and *swive*. Both can be used as either a verb or a noun. (175)

The terms are vulgar and anachronistic. They could have no value unless generally adopted, and there is no reason to assume they would be. I recall a friend in high school who prided himself on his erudition. He used both *quim* and *swive,* among other archaic words, often those with a sexual or scatological meaning. Needless to say, they did what one might expect: marked him as unfamiliar, peculiar, and particular. Most would not call *quim* and *swive* "jargon," but they fit the *OED* description. So why would Money waste his time with this? And the time of others: he acknowledged that "Elizabeth Connor, of the Welch Medical Library, gave generous time and effort to lexigraphic searching" (1982b, 173).

It might seem somewhat surprising, but Money's work with language had the same intention as his work with pornography. The emphasis on dirty words reflected Money's concerns for communications, the concerns that led him to appear in a wide variety of magazines and newspapers. He wished the ideas to get to anyone who would listen and to transmit them in a form that would be understood. No doubt it also appealed to his theatricality to have his dignified professional demeanour contradicted by the vulgarities. The sound of *quim* and *swive* would follow nicely on the word *fuck*. But then how does this fit with words such as *autoagonistophilia* (pleasure from being viewed while having sex)? This seems the other side of that liberation, the need for scientific independence and precision. Each new word would make the understanding of a condition more exact. Money prided himself on the breadth of his knowledge, in a variety of directions, and so the Greek and Latin creations appealed to him.

Still, the contradictions are probably not ultimately explainable. In one of his various articles on the universal exigencies, he wrote,

> In recognition of the fact that psychology's units of raw data all serve to *indicate* something or other to the psychologist as scientist, they can all be categorized as *indicatrons*. The principle under which they are subsumed can go by the name of *indicatronics*. (1974b; 1986d, 111)

He then proceeded to use the words *behavioron, imageron, verbaton, dictaton, graphicon, praxicon,* and *systematron* (1974b; 1986d, 113). As for Adam at the beginning of the world, there were so many new things that needed naming, and God seemed to have given the job to the atheist John Money. Sadly or happily, the rest of the world has seldom converted to his religion of neo-onomastics.

CHAPTER SEVEN

What Wild Ecstasy

Sexual Liberation, Pedophilia, and Pornography

John Heidenry's *What Wild Ecstasy: The Rise and Fall of the Sexual Revolution* (1997) uses many sources, including sexologists such as Money, but his focus is more on pop celebrities such as *Penthouse* publisher Bob Guccione and *Playboy* founder Hugh Hefner. I wouldn't go so far as to call John Money the Hugh Hefner of sexology, but there is a certain similarity in the way they fit the age. Most sexologists are at least slightly out of tune with their society. A prime example would be Kinsey himself, who was too conservative in his relationship with society to be comfortable with the quite radical sexual knowledge his scholarship was discovering. Then his own move into more radical sexual practices was professionally and personally disastrous, as revealed in the biographical film, *Kinsey* (Condon 2004).

In contrast, Money's professional exploration of sex and personal exploration of sex took place in the 1960s and '70s, when both adventures suited the culture. Money had very little personal life away from sexology, so he was in a perfect position to let both professional and personal expand as they willed. Money was the man for the zeitgeist. At times, Johns Hopkins seemed at least slightly proud of Money, but it returned to being a stern parent well before society did. In an interview in *Pillow Talk* in 1981, Money had the following exchange:

PT: You have been called the "guru of sexuality." Does the title bother you?

MONEY: If I was living in a hermitage, I couldn't care less. But in the conservative Medical School of Johns Hopkins, I'm not sure they appreciate having a person known in *Baltimore Magazine* as their resident "guru of sexuality." But what's in a name? (1981b, 87)

In *The Riddle of Gender* (2005), Deborah Rudacille interviewed Paul McHugh, head of psychiatry at Johns Hopkins when the Gender Identity Clinic was closed. McHugh said that Money was "a victim of the sixties":

> His science brought him so far, was bringing him so far, and then, like so many other people, the theme of "overthrow the patriarchy, make change, it's the authority structures that are standing in our way" – I think that John bought that hook, line, and sinker, and as a result, like so many others, came to suffer from it. (2005, 124)

Yet the less establishment aspects of society continued to respond to Money's message. *Rolling Stone* named Money "Hot Love Doctor" in its "Hot List of 1990" (1990). Then, as the zeitgeist as a whole shifted in a conservative direction, Money was both hurt and angry. The progressivism that had become the core of all aspects of his life was under siege from every direction.

There are many examples of Money rocking the boat in his early years at Johns Hopkins. A number of them come from the 1950s and early '60s, when he might be called a premature sexual liberationist. In 1955, he wrote "A Children's Medical Book of Sex," which included the following: "Many people have asked me what is the suitable age at which a child should be given this book to read. The answer is that any child who is competent enough to read the book, or part of it, is old enough to know the information he will read" (1955a, 4). He submitted it to a number of publishers. None accepted it. The rejection letter from one publisher is representative of responses to the manuscript:

> Two of us have read your manuscript with admiration for the way in which you have handled this difficult subject matter. I hope I am all wrong but I think the market cannot take as frank and factual a presentation as you offer. Verbal frankness is one thing; material in book form is quite another. (1955a, 19)

As one might expect from his various comments about sex rehearsal play, Money believed very strongly in early education in sex. The value of pornography was a common subject in his various talks on sexual education, and at a 1971 seminar on sex in childhood the title of his paper was "Pornography in the Home" (later published as a book). There he said, "One of the reasons why I'm interested in bringing the whole subject of pornography and sex out into the open in the family is because I would like for all children to be turned on the most by normal sexual fantasies and not by the 'screwball' ones" (1971c, 78). He ended the talk with one of his favourite examples, a Sunday School teacher describing Christ's death. He asserted that, although the teacher explains "how to commit a crucifixion," Money had yet to hear of "children who have come home from Sunday school and played crucifixion games, to crucify their dolls or their playmates" (83).

Money often commented in different ways on society's obsession with protecting children from the knowledge that he thought essential for development. He saw Christianity as part of a general philosophy that allowed the depiction of violence but not sex:

> The best a good social anthropologist can say is that we are a society that for all its recorded history has been death-loving. We have a religion that is formulated on the basic principle of death, and killing and violence are perfectly acceptable on the front covers of magazines. (1977k, 56)

Money had no ongoing interest in Freud, so there seems to be no reason to infer that he was saying anything about the pleasure principle and the death drive. Instead, he was considering what have often been called the two guiding forces of popular culture, sex and violence. Money was far from alone in noting that censorship is much more concerned with hiding sex than violence. It is a commonplace in discussions of censorship and particularly in the age markers of film classification; children are allowed to see graphic depictions of violence but very little overt sexual activity.

He began a talk to medical students at the University of Maryland in 1969 with a typical Money phrase to jolt them out of complacency: "I think we might even reverse our traditional attitudes and say, if I may be slightly shocking, that it would be good to have 'fucking' games in kindergarten" (1969b). He then admitted he was exaggerating but turned to a more careful and nuanced version of the same:

> Love education is also a thing that begins in early childhood because if you pay attention to kids in kindergarten you will find that a remarkable number of them have love affairs ... I think these kindergarten love affairs which are genuine rehearsals of romance, rather than sex itself, ... fade out after this early age and in middle childhood you don't get too much of it ... They have little secret love affairs ... I have concluded that children must be extremely variable with regard to this period of life and the amount of sex life that they have. Certainly I think under ordinary conditions of sufficient freedom all children in their middle childhood years are liable some time or another to play sex games ... Perhaps it is really true that this is the period of life when you are so engaged in consolidating your own sense of gender identity and gender role that it is necessary for boys and girls to separate themselves pretty strongly, especially as far as the business of love affairs. When they come together, it is for a brief episode of possible sexual play which proves to them the correctness of their gender identity and gender role. It would be dangerous, perhaps, for girls to play with boys or boys with girls all the time, constantly, because one might become too much like the others. The big assignment in life at this time is to become different and separate in gender identity. (1969b)

For Money, "sex education" is about "love education," and the two are intertwined, as he showed in his theory of the lovemap. He saw the "love affairs" – and no doubt he would include wedding games – as part of this but also as part of the consolidation of gender. Money often asserted that he did not see humans as innately defined by being male and female, yet he also recognized that the whole history of human society is devoted to making people be one or the other. The small love affairs affirmed the complementarity of gender, that the self is one gender and that the other is the other. The conclusion of this quotation is about the necessity of this gender division; it is also about how tenuous and insecure it is. If gender identity is not innate, then it needs reinforcing. Perhaps heterosexuality does as well. Early gender division enhances this as does early sexual experience. Money was very aware that the world wanted heterosexuality to be the norm, and he believed that early sex education and even sexual activity could be seen as a means to this end. The mores of society expected heterosexuality, and the way to achieve this was to let children explore their developing sexuality and developing sense of gender in a manner free from the restrictions of this overly moralistic society. Today the majority of North Americans would be offended by an argument for sex education based on the suggestion that it would decrease homosexuality. Yet, regardless of this change from the attitudes of the 1970s, explicit sexual education remains an uncomfortable idea. If he were speaking today, Money might say instead that the obvious appeal of sexual education is that it improves psychosexual function, whatever the orientation of the individual. Ultimately, his kind of explicit universal sex education from a very early age might lead to vast changes in the balance of the continuum of sexual orientation or no change at all. Regardless of what such education might create, it is unlikely to be attempted.

In the late 1960s, Money began to offer a course at Johns Hopkins in sexual medicine, which included a lecture he called "Pornography in the Home and in Medical Education." The course and the talk quickly became quite well known, with no little help from Money's own self-promotion. He often spoke on pornography to community groups and to academics. *Medical World News* (1971) published an article about the course titled "Med-School 'Stags': Students Learn about Sex from Pornographic Films": "The noted sex researcher told the recent American Psychopathological Association meeting in New York City that the course, in presenting an extreme range of sexual material available in today's society, encourages students to re-examine their attitudes towards sexual mores and to develop an objectivity that will help them to deal with patients with sexual problems." The idea received significant praise. Dr. George Andrews (1971) wrote a letter in reaction to a *Medical Tribune* article about the lecture:

> Familiarity with the best available pornography would seem desirable for all people doing counseling or psychotherapy so that everyone (as well as M.D.'s) would be able to get speedily over the anxieties, implicit or explicit moral judgments, defensiveness and the like, that occur in response to patients' reports of the sexual aspects of their lives not approved of or unfamiliar to the therapist.

But it also received many attacks, including one from Charles Socarides, who wrote, "I have completed a major research paper on this issue derived from twenty years of teaching and treatment in the field of sexual perversion" (1971). Socarides claimed, "Recourse to such materials will induce the inexperienced to seek and to expect similar situations." He concluded, "Any recommendation that pornography could benefit instructional programs for medical students or could prove benign in the sex education of anyone unquestionably reflects a complete lack of clinical judgment." In a handwritten note, apparently not sent, Money responded, "Dr. Socarides speaks not as a scientist, but a dogmatist."

In 1972, Money wrote a letter to an English doctor, Michael Simpson, to explain his position on pornography in medical education. The range of his hopes makes it worth quoting at length:

> I use the topic of pornography to open up the issue of the visual image in human sexuality, in both its normal and paraphiliac manifestations. At the same time by using commercial pornographic illustrations, it is possible to integrate the topic with medical sociology. One purpose of this lecture is to put a metaphorical egg-beater in the brains of the audience, so that they will be able to reformulate their attitudes on sex, with particular attention to the physician's position of being nonjudgmental about all matters of sex that his patients may bring him. I lay particular emphasis on the development of an attitude of nonjudgmentalism toward patients. I ask students and others to bring a partner of the opposite sex, so that it becomes possible to establish an exchange of ideas between the sexes, which is particularly pertinent since the visual images seem to serve a somewhat different function in males and females. I find that women in general are grateful to be at last not excluded from the affairs of the "men's house." (1972c)

There are many elements here representative of Money's attempts to be a proselytizer of sex education. He saw commercial pornography as not debased but a cheaply acquired asset: even in the 1960s, resources in medical education were expensive, but explicit pornography was not. The "metaphorical egg-beater" was always in Money's hands, as he thought almost everyone should be compelled to be more open about sexuality. Money emphasized "nonjudgmentalism" first for professional reasons, so each clinician in sexuality could help any patient, regardless of sexual practice, but also for the general public, to move towards a society in which many paraphilias, of little attraction to the majority but of little harm either, would be accepted by the general populace. This was part of the purpose in the last comment. While Money wished to see more women in medical professions, even more he wanted women to be more knowledgeable about and open to their own sexualities. Given that the vast majority of paraphilias that he encountered were limited to males, he also needed knowledge and nonjudgmentalism from the women who were very unlikely to be attracted to such practices.

In a less professional vein, Money had asked to use some empty space at Johns Hopkins to promote his developing knowledge of contemporary art. In 1963, he wrote a letter to his friend Isadore Rubin, editor of *Sexology* magazine, about the problems of censorship:

> I have run into my own problems of pictures of the sex organs today, with a complaint from a medical student about the art exhibition that I hung, as per routine, in the Medical Residence Hall this weekend ... The priggy student is worried about bringing his father and mother in contact with the pictures! (1963d)

Money's connection with Rubin was typical of his general outreach project. *Sexology* began in the 1930s, one of the many magazines created by Hugo Gernsback, more famous for his promotion of science fiction, but it had its most significant presence in the 1960s. Money often wrote polemical articles for *Sexology* and for other pro-sex magazines that flirted with pornography, most prominently *Forum*. Peter Luce, the character in *Middlesex* that Jeffrey Eugenides modelled on Money, has a column in *Playboy* titled "The Oracular Vulva," in which the titular genital explains female sexuality to the male readers (Eugenides 2003). Money never quite went that far, but he did appear in *Playboy*'s own "Forum" in the July 1990 issue.

It began with a note from Hugh Hefner to the "Forum" staff:

> In an excellent piece on Dr. John Money in *The New York Times*, dated January 23, 1990, Dr. Money predicts that "current repressive attitudes toward sex will breed an ever-widening epidemic of aberrant sexual behavior." The *Times* story goes on to say, "As for malignant social influences, he and other researchers found no evidence that pornography causes or fosters the development or expression of paraphilias [sexual perversions]" ... I would like to suggest that *Playboy* find a way to explore some of the issues raised by Dr. Money's research. (1990e, 45)

The interview that followed is titled "Sex: The Good, the Bad, and the Kinky," and the subheading states, "one of America's premiere sex researchers takes on the forces that want to undo the sexual revolution." The introduction refers to the "wide-ranging conversation. We found his outspoken views to be provocative and dead-on" (46). The comments are a condensed version of Money's major ideas about the lovemap, but the primary target is what he called "a sexual counterreformation," a response to what the popular press continues to call the "sexual revolution" but that Money described as the "sexual reformation." The essence of his polemic appears in two sentences at the beginning: "Essentially, some people are taking everything that is sex-positive and labeling it sex-negative. Today's witch-hunt goes after women's liberation, gay liberation, sex education, contraception, teenage pregnancy, abortion and pornography" (46).

For Money, it was not just the unfairness of the witch-hunt but also the evil results he perceived as an inevitable result of such repression. In the *Johns Hopkins News-Letter* in 1977, in "Doctor Money Speaks, the Whole World Listens," Money referred to the hypocrisy of contemporary society: "For sex the message is that it's all right as long as you keep it secret – I could say dirty, underground, except then you say that it's sacred and you just keep it for the one you love" (1977e, 1). Money believed the effect of such secrecy had already been demonstrated in other contexts: "Censorship just leads to 'the gangsters.' We did that with Prohibition. We're doing it with marijuana and next we're going to do it probably with sex. It's been that way with prostitution for practically as far back as you can remember" (2).

Money made it clear in the *Playboy* interview that he believed the sources of the new sexual freedom were quite simply penicillin and the pill. In other words, he saw an openness to recreational sex as an inevitable product of scientific progress and specifically progress in medical treatment that focused on the potential negative results of sexual activity. In 1973, he wrote in the *Chronicle of Higher Education,*

> In brief, the invention of effective contraception has coincided with other factors that govern population dynamics that make it not only possible, but also imperative, to differentiate recreational from procreational sex. Earlier in this century there was great resistance in our society to this heretical idea. Change is always threatening, especially where a taboo is concerned, and sex is our maximum taboo. Resistance to the idea of recreational sex began crumbling first within the confines of the institution of marriage. The wall of resistance has still not fallen completely, but there is now wide acceptance of the proposition that married persons ought to have most of their

> sex for fun, deliberately restricting the possibility of procreation and limiting the size of their family. (1973b, 20)

The general acceptance of contraception had coincided with – and no doubt contributed to – the popular belief that an ardent sexual life was the key to a happy marriage. Today there are many books that focus on how to combine a lively marital sexuality with a very conservative religious belief, but the "institution of marriage" remains absolute in these books, and sex is seldom described in as simple a word as *fun*. The books listed on the website themarriagebed.com show that Christian sex has become a growth industry, but always marital sex. Money saw the natural progress of society as leading away from any of these restrictions. If the only legitimate inhibitions to sexual liberation were infection and reproduction, then the introduction of any other limitation was self-evidently wrongheaded. Instead, he foresaw a progressivism that was almost boundless, as he claimed in *Forum* in 1982:

> The consequences of this contraceptive revolution are as profound and far-reaching in human affairs as were the consequences of the discovery of how to make fire.
>
> The fire-making revolution applied not only to domestic cooking and heating, agricultural and hunting burn-offs, and fighting, but also to the divine flame of religious worship on altars of burnt offerings.
>
> The contraceptive revolution, by contrast, has not yet been applied to or incorporated into religious worship. Rather, religion is still in the throes of ideological adjustment to the new era of contraception and the freedom and sexual reformation that it has sparked. (1982c; 1986d, 517)

Money was interested in sexual practices of all cultures, but he had little knowledge of the sexual theories of non-Western religions, even as many of those theories began to become a significant part of Western sexual liberation. For example, I have found no reference in his writings to tantric sex. As in so many other aspects of his cultural awareness, he thought in terms of scientific liberation and religious repression, primarily Christian, as in *The Destroying Angel*, his 1985 attack on the history of the anti-sex movement in America. In that book, Money also gave credit to the progress offered by technology:

> Today the remnant of degeneracy theory, otherwise known as depravity theory, is quietly being discredited by new video technology. The new technology allows more and more people to see erotic and pornographic material, privately at home, and to discover that they do not become depraved from it. (1985b, 174)

In a sense, Money's public arguments were always about the lovemap, about providing the sexual education that enabled all individuals to develop the most successful sexual function that was within their power. In "Significant Aspects of Erotica," Money was able to make quite a radical argument seem like simple common sense:

> Historically, a significant majority of the general public subscribes to the idea that erotica is fundamentally dirty and should be forbidden. Adults are prone to accept these notions and quite intentionally, as well as obliquely, to teach their children that exposure to either visual or narrative portrayals of eroticism is evil and sinful. As a point of departure for sound psychosexual development, this rationale leaves much to be desired. (1970h, 1)

He concluded, "Obviously, the picture of normal heterosexual intercourse is a laudable representation of the most desirable form of human sexual expression" (9). Our fear of children seeing an activity that we consider to be a healthy part of everyone's life does seem like an unusual exception to our usual practice. I cannot think of anything else that we regard as completely positive but keep from children. Just as one example, we might keep children from seeing violence that we believe necessary to keep civil order, but we see that violence as a necessary negative rather than "completely positive."

Money's goal was sexual education in the widest possible sense, as his speech to the National Commission on AIDS in 1992 demonstrated: "My secular heresy is that we should embark on a nationwide program, equivalent in magnitude to the Manhattan Project or the Marshall Plan, to eradicate antisexualism from our midst and to replace it with a positive philosophy, or sexosophy to use the term I prefer, of prosexualism" (1992b, 20). Hitting his stride, he had a still greater vision, something presumably much more educational than the Playboy Channel or Naked News: "There will be special television channels designated as specifically sexual and erotic. They will combine sexual news and information with erotic entertainment of the highest caliber" (20). In this world, medicine will be transformed:

> Even for prepubertal juveniles there will be none of the euphemizing and evasiveness that is today so complete that, all across the nation, there is not even one clinic that specializes in pediatric sexology and sexual health. Amazingly the same applies to adolescent or ephebiatric sexology and sexual health. (20)

Lest it might be thought that Money's views were popular with only an academic audience, it is interesting to note that the speech was reprinted in the *Kansas Biology Teacher*.

Money saw himself as a leader in the charge towards sexual liberation and wrote accordingly. Thus, in 1978, he wrote for *Forum* "The Sexual Revolution: A Manifesto." Money was so often seen as avant-garde and arrogant, the personification of the elitist who knew so much better than the general populace, but his manifesto had a very different tone, perhaps the one that appealed to that Kansas educator:

> We in America do not live in a sexual democracy, but in a sexual tyranny. Politically that is very dangerous, as all tyranny is dangerous. Dictators, including Hitler and Stalin, have always used sexual tyranny as a means to total tyranny. After people have bowed under the weight of sexual tyranny, they often bow under the weight of total tyranny, mistaking their martyrdom for righteousness. (1978g; 1986d, 513)

Rather than Money being the communist devil depicted by some of his opponents, it seems that he was a devout democrat of the sort so revered in the American tradition. This was his justification for many of his public endeavours, such as being an expert witness in the obscenity trial of *Deep Throat*, arguably one of the first pornographic films to enter the mainstream imagination of the United States. Money testified that the prosecution failed to recognize the general change in attitude in the United States:

> It is if you have your thing that you like to do and you mind your own business with it, it doesn't offend me, that is the attitude I hear over and over again. People are allowing pluralistic possibility for other people's sex lives and not trying to force them all down the same runway as used to be the case. (1973c, 272)

He justified the film on the basis of its accuracy: "By and large they were in that film not faking, they had some kind of relationship existing, some kind of genuineness of what was happening between them" (273). When the prosecution presented a comment by an NYU psychologist to the effect that *Deep Throat* reduced humans to animals, the defence counsel asked for a response:

> KASSNER (defense): "As a psychologist, what would you say about the position of a person who writes about sex using the words 'animal activity,' 'animal coupling,' 'animal connection' in depicting the human activity of love making?"
>
> JM: "He has an impoverished sex life of his own." (279)

Money was never above using ad hominem arguments against his opponents.

He often spoke out against legal restrictions. The aversion to the law that seemed evident in his expert testimony was only partly a reflection of his liberationist ideology. He saw the law as impeding both social progress and legitimate medical choices. Yet, as a specialist in various sexual deviances, he was a part of the legal system and wrote frequently on what he called "forensic sexology." In one such piece in 1983, in an editorial in the then new journal *Medicine and the Law,* he wrote,

> Society claims its right to intervene, though not to treat or prevent, but to exorcise by imprisonment or death. One of the wretched consequences is that society withholds its money from research into the cause of sex-offending, thus denying itself the possibility of prevention. (1983c, 158)

Money was of course not alone in seeing the medical model of deviance as in conflict with the criminal model. Every argument about an insanity defence is based on that conflict. In an article in the same volume, he and Greg Lehne called for a change in the attitude shown towards sex offenders:

> The divergent assumptions of the sexological and the legal model of human erotic and sexual behavior on the issue of voluntary choice and causality limits the ultimate convergence of the sexological and the criminal-investigative understanding of the paraphilias. The amount of convergence presently evident applies to the criminal-investigative phase of the criminal justice system and not to the adversarial-judicial and the judgmental-dispositional phases. (1983b, 261)

The language here is a bit dense, but the argument about causality is central to Money's view of sexuality. While he often explored arguments about causality, he found them useful primarily in support of theories that might help to prevent paraphilias in later patients. He recognized that there was some degree of choice in all human acts, but he also saw that judging whether a person has the ability to prevent making a certain choice is usually difficult or even impossible. Instead, following a medical rather than a criminal model, the response of authorities should be to pursue not punishment but methods that enable the patient to avoid the choices that the authorities deem criminal and pursue the choices that the authorities deem necessary.

Thus, the impetus for Money's sexual liberation was always both democratic within a very American individualist tradition, to allow sexual expression as long as it didn't scare the horses, and scientific, to allow both all efficacious medical treatment and all potentially useful medical research. Money saw his primary opponent as the moral outrage he noted in *The Destroying Angel.* In one anti-gay video in 1992, a speaker claimed that "Dr. Money from Johns Hopkins University demands much more than just legalization of adult child sex ... He argues if boys die from consensual sadomasochistic activity, adult killers should not be prosecuted" (cited in Meehan 1992). A reporter asked him for a response: "Money was flabbergasted when asked about the video. 'I have absolutely no idea what they are up to in making that statement ... It doesn't belong to me'" (cited in Meehan 1992).

Money no doubt was very sure "what they [we]re up to": it was the public face of an evil "sexosophy": "Sexosophy is defined as that body of knowledge that comprises the philosophy, principles and knowledge that people have about their own personally experienced erotosexuality and that of other people, singly and collectively" (1981d; 1986d, 569). He always used the term in opposition to sexology, "the science of sex" (569). He seems to have intended sexosophy to be value-neutral in itself, but that 1992 promotion of a sexosophy of "prosexualism" noted above was a rare positive use. To put it in Orwellian terms: sexology = good; sexosophy = bad. Thus, in 1983 he wrote,

> Victorian sexosophy still exerts its effect on sexual psychoneuroendocrinology. The neuroendocrine part of this conjoined discipline is now scientifically respectable, whereas the psycho-behavioral part is not, at least in human investigation. The latter is poorly funded. It is still stigmatizable as obscene and degenerate. To be too closely

> associated with it depletes, if not one's vital spirits, then one's professional status. (1983d, 399)

Money's view of the evils of "Victorian sexosophy" and his emphatically positive view of sexual expression preclude much recognition of the sexual attraction of negative responses. As Foucault (1990) and many others have pointed out, Victorian repression was but one side of a Victorian obsession with sex, often expressed in desire for sexuality represented as disgusting. Freud often noted the attraction of disgust for many people (2000). As in Money's view of the dysfunctionality of a lovemap in which love and lust were divided, he had difficulty seeing the positive in the negative. His sexual liberation had a tendency towards sexual libertarianism; thus, there was little room for the joys of repression. At the same time as he was accepting of paraphilias such as bondage and discipline, he did not easily accept the erotics of repression as a more general principle.

Arguably, the suppression of medical freedom has no such obverse. Money often warned his graduate students and postdoctoral fellows about that depletion of professional status. In 1962, he already recognized the dangers of "the risky frontiers of science" (1962a; 1986d, 499) and called for independent thinking in spite of risk: "The sex scientist may follow another way in making a truce with antagonistic popular values. He may approach a touchy topic, carrying his investigation only to the limit of what is socially tolerable" (1962a; 1986d, 497). But "some intrepid souls will come forward, as did Galileo and the rest, to challenge society and its personal values, when there is promise of new harvest from the tree of sexual knowledge" (1962a; 1986d, 499). Money usually saw any interference with sexual research as unacceptable. In 1981, when asked to write a futuristic piece on "Male Sexuality in 1990," his emphasis throughout was on how taboos would inhibit the potential for scientific development: "Sexology does not have academic and scientific status and respectability. Urology does: it is respectably urinary, except that, by prudish misfortune, the urinary system shares joint tenancy with the reproductive system" (1981c, 53). In 1977, Money wrote a piece that admitted the evils of past genetics research but suggested an equal evil was then being perpetrated in the prevention of genetics research:

> Restriction of cytogenetic screening signifies to me anti-science and anti-therapeutics. The pendulum of the ethics of informed consent has swung too far. Instead of being protected, patients are being

> exposed to the risk of missed diagnosis or wrong diagnosis, lack of treatment or wrong treatment, and a bad prognosis. (1977c, 130)

It might seem that Money was just being self-serving – the researcher thwarted. Still, he was never only a researcher: he had a strong belief in the efficacy of responsible, knowledgeable treatment. His obsession with longitudinal studies, and with finding a way to keep patients within those studies, was ever intertwined with his clinical work: the one inevitably supported the other, and both were of value first to the patient.

In cytogenetic screening, the harvest came from the seeds, but there was a possibly greater benefit from what might be called the young plants. For Money, children were first the source of the lovemap, the place that a functional or dysfunctional sexuality developed. He repeated like a broken record that the best way of stopping sexual abusers was to prevent those abusers from being abused as children. This has now become the accepted theory, yet in 2014 we do not seem any better at preventing the originating abuse than when Money was speaking fifty years ago. The twofold answer to the problem for Money was to improve sexual education for children, as noted, and to study the sexuality of children:

> Science also has become obligingly subservient to the insatiable power of taboos – so much so that scientists have in recent years invented the bastard science of sexual victimology in order to get governmental or private research grants. Today, sexual research is funded only when it defines sex as a crime, focuses on the alleged victim and excludes the alleged offender, whose fate is under the jurisdiction of the law, not sexual science and medicine. The offender is very often incarcerated, as epileptics once used to be.
>
> There is no research funding for sex as a positive experience, for falling in love and pair-bonding, for lovesickness or for the sexual rehearsal play of childhood. I cannot conceive of any researcher's submitting a grant application to study juvenile love affairs, or juvenile lust and intercourse! (1982c, 43)

The term *victimology* seems to have been first used by criminologist Benjamin Mendelsohn (1963), sometime during the Second World War, but it has since become a recognizable field of study, and, like any field of study, it might be opposed for one reason or another. But Money seems to have been blaming all victims of everything and blaming all who help all victims, which seems more than slightly extreme. Still, there is no question that the law has become more and more focused on the victims of crimes. It is interesting that Money published this article the same year the "victim impact statement" became a part of law in the United States (Travis 2008, 518). Since then, the victim impact statement has become normal and is part of a common assumption that the question of legal culpability is less important than the effect on a person who has been injured in some way. The result of this process has been to give even less support to those who would emphasize medical care for the person who perpetrated the injury.

"Sex as a positive experience" is even more beyond the pale. Many researchers have noted the necessity of tying sex research to the prevention of sexually transmitted infections so that epidemiology and innocent victims of contagion can be presented in applications for funding. And then there is Money's call for research into juvenile sexuality. At the same time as society constantly questions the possibility of an innocent adolescent, all social media and all law enforcement agencies assert the importance of maintaining the innocence of adolescence. There have been many stories about how easy it is to get funding to catch "Internet predators," as opposed to getting financial support to do anything for the sexual education of those on whom the predators are seen to be preying. To educate the victim to not be a victim seems of far less interest.

While Money wrote most often about sexual education for the young, he also asserted the importance of being able to examine positive sex as practised in all aspects of society. In 1967, he wrote "Sexual Problems of the Chronically Ill," in which he noted the near impossibility of treating such problems with only anecdotal evidence:

> Problems of sexual functioning are still subject to a strong taboo against being demonstrated in action with the partner. The doctor must be guided only by verbal reports. No other specialty in medicine thus denies itself a first-hand clinical examination of the nature of the patient's complaint, to say nothing of the elaborate laboratory apparatus used to probe more deeply than the naked senses can perceive. The advance of medical knowledge will surely demand the abolition of this taboo both in research and in diagnosis and treatment. The first beginnings have been made and published in the now famous report of Masters and Johnson. The public is probably ahead of the profession in its acceptance of whatever in science and medicine

> will benefit this highly treasured function of everyone's life – sex. (1967c, 286)

Given that Money wrote this just after the publication of Masters and Johnson's first book, *Human Sexual Response* (1966), based on examination of physiological response during sexual activity, it is easy to see why Money was pushing for "first-hand clinical examination" and "elaborate laboratory apparatus." He often used *taboo* as the term is commonly employed rhetorically to represent some proscription against a certain practice as though that proscription were a superstitious holdover from some primitive religious belief. I find it somewhat difficult to understand Money's view of "the public" here, however. It could be just the typical political comment: "Everyone seems opposed to me, but the silent majority is with me." Money had more than enough experience with a general public that seldom supported his calls for a wide acceptance of sexual possibilities.

Early in his career, the degree of freedom Money believed he had in the exploration of adolescent sexuality seems to have been extraordinary, judging by two transcripts in his papers. There is no date on the first, but the second is marked 1970. They are transcripts of two interviews with four teenage boys, two fifteen years old and two sixteen years old. The boys are all named, although there is no note as to whether these could be pseudonyms. One particular question in the first transcript suggests the kind of direct interrogation pursued during the interviews: "George (15) WHEN YOU'RE LOOKING AT THEM, WHAT DO THEY DO TO YOU? Git me a hard on" (1970i, 2). The second transcript concludes, "I'M PUTTING THIS TAPE BACK ON CAUSE I WANT TO GET THE POINT CLEAR THAT BOTH OF YOU TRY TO PULL OUT SO YOU DON'T GET THE GIRL PREGNANT RIGHT? Yeah. O.K." (4). It is difficult to see how Money could have justified this. Regardless of his personal views on the topic, he must have realized that almost any medical researcher would have seen these interviews, with no suggestion of parental consent, to be unethical. I have no idea whether Johns Hopkins knew he was doing this, but my guess would be that it did not. While these interviews represent Money's belief in sexual research with children and the importance of the researcher maintaining a nonjudgmental attitude, they certainly would offend anyone who either believed in the innocence of children or believed that professionals should do nothing to create any suggestion that non-innocent behaviour of children would be acceptable. The interviews also reveal the arrogance that so often caused Money trouble and was used against him in Colapinto's *As Nature Made Him*.

On the other hand, Money's treatment of these teenagers, who, judging by attitude and dialect were probably uneducated street kids, shows in a wonderful way his refusal to treat such people as either criminals or victims. Victimology in sexology tends to support an attitude that the focus of the study of any sexually deviant act should be not the perpetrator but the object and that the object should be invariably seen as object, without agency, and being an object should be considered primarily as the victim of what a perpetrator has done. Money's rejection of the assumption of who is a victim, and his belief in the necessity of a child's sexual development, led him to take a position on pedophiles that was and is in direct conflict with the general populace. In an interview in 1986 with *Omni* magazine, Money quite nonchalantly explained how a pedophile might operate, and the interviewer responded, "It's a little too complex for me." Money replied,

> Well, I'm not surprised that it's complex, because, you know, the average person who works in child abuse and sexual abuse and victimology just simply doesn't learn all of these things because they're so carried away with playing St. George slaying the dragon that they never talk to their patients. Well, of course, if they see the child whom they label as victim, even though the kid be a very positive participant in it and doesn't think of himself as a victim at all, they still don't even go and bother and talk to the older lover, so they miss all of these things. (1986a, 82)

In Money's papers is an unpublished manuscript dated 1978 and titled "Pedophiliac and Epheboliac Relationships: A Pilot Study." It included the following paragraph:

> In some instances, the patients have entered into a pair-bonded relationship with younger partners. In these instances the relationship between partners has been reciprocal and complementary. The younger partner, of course, conforms to the imagery of the erotic "turn-on" of the adult partner. Not always obvious, from cursory observation, is that the relationship may have a payoff in pair-bonding, typically economic and recreational, as well as affectionate and erotic for the younger partner. In the younger partner, the history may be one of parental deprivation and neglect, intentional or nonintentional. (1978f)

The typescript concluded:

> The relationship between an older partner and prepubertal or early pubertal boy may be: (1) complementary and reciprocal; (2) affectionate and pair-bonded; (3) exclusive of assault, harm or injury; (4) time limited; and (5) phenomenologically a sequel to child deprivation, neglect, or abuse. In such a relationship, the ethical and legal issues of the sexual rights of childhood and adolescence need to be re-examined and redefined. (1978f)

What Money called "the sexual rights of childhood" have seldom been re-examined and certainly have not been defined. His idea that a pedophile relationship could be better for the child than the parental one was out of keeping with the beliefs of his age and continues to be in ours. In 2004, Steven Angelides published an article with the telling title, "Feminism, Child Sexual Abuse, and the Erasure of Child Sexuality." He concluded, " In a contemporary context of escalating anxiety and panic surrounding pedophilia and child sexual abuse, it is increasingly difficult, and perhaps for this reason all the more imperative, for queer studies to problematize the cultural and relational construction of age, child sexuality, and subjectivity" (168). The language is very different from Money's, but the sentiment is the same – to acknowledge the sexual agency of the child. The impossibility of pursuing the research to support such studies is suggested by the title of *Censoring Sex Research: The Debate over Male Intergenerational Relations* (Hubbard and Verstraete 2013).

Another human that Money wished to save from victimhood was the adult woman. Yet many would hesitate to see him as a feminist. The explanation for his claim to feminism and for the attacks on him by feminists is visible in a review he wrote in 1977:

> The old ethic was duple: by one criterion, recreational sex was illegitimate; by the other, the evidence was to be kept hidden, restricted to the locker-room, stag party, or whorehouse, and was preponderantly for men only. The developing, new ethic is single: sexual recreation is not sex segregated, and it is less covert. In other words, some of the taboo against sex is lifted. Though the ultimate repercussions of lifting a taboo in society are not well understood, breaking or easing the taboo upsets, indeed enrages many of those who obey it. (1977m, 176)

Money believed sexual freedom to be key to women's liberation. In the soft-porn magazine *Gallery*, he said, "I've talked to many women right here in this office who still have extraordinary difficulty accepting the fact that really nice, high-class, good-thinking women enjoy fucking" (1977k, 57). Money always asserted that the markers of gender are culturally assigned with the exception of reproduction. Thus, a successful liberation for woman qua woman must begin with the sexual. Money, however, would be seen by many in the same light as other male sexual liberationists – not seeking liberation for women from sexism but seeking to free women for the use of sexual opportunists. Women who were truly "good-thinking" would recognize that it is the men who "enjoy fucking."

By the time *Gendermaps* was published almost twenty years later in 1995, Money had refined his understanding in a typically bold and complex argument:

> To some extent, waning enthusiasm for mainstream feminism on the part of younger women in the early 1990s may be a concomitant of failure to address the apparent contradiction between equality of men and women on the dual criteria of gender and lust. Rather than sameness between men and women on the criterion of lust, there is reciprocity. The male-disparaging crusade of radical feminists helped by counterreformationists has become too strident, perhaps, for the generation of the nineties. Insofar as the infidels in this crusade are men, women may be held accountable for its stridency. Ironically, however, some of the crusaders are men. The secretive and self-righteous purpose of these male crusaders is to restore the prefeminist status quo by sacrificing some of their own sex in order to demonstrate that women are imperiled by men other than themselves. Simultaneously, this demonstrates that women are unable to be independent and autonomous, and must therefore be brought under the dependent protection of the likes of themselves. Scarlett O'Hara must live again!
>
> Feminists who were once blinded by this ruse and who now see through it dissociate themselves from man-bashing tactics. The ultimate absurdity toward which man-bashing tactics lead is that of the disposable male, namely that, since only a few stud males will suffice for breeding, society could dispose of the rest. (81).

The author of this article showed none of the anti-male bias attributed to John Money by *As Nature Made Him*. On the other hand, it might seem strange for someone to be crusading against "man-bashing" in 1995, when much of the truly anti-male feminism had long gone out of style. Anyone who enters a gender studies classroom today, however, will find many female students who reject the label "feminist" because they associate it with a "man-bashing" that was no longer part of feminist rhetoric before they were born. The residue is clearly there. The reason for this, no doubt, is multifaceted, but part of it is probably exactly as Money said, the failure of feminism to embrace an active "lust." While the lesbian feminists of the 1970s often rejected an ardent sexuality, those of today have no trouble praising recreational sex. Heterosexual women, however, for the most part seem to believe that an ardent pursuit of sex precludes identifying as a feminist. Those who label themselves as some version of "sex-positive feminists" are very much exceptions, and the most prominent ones, such as Susie Bright (a.k.a. Susie Sexpert), tend to identify as bisexual, which in itself takes them out of the mainstream.

While Money was always clear that gender is not biologically defined, neither can gender be divorced from sex: "Gender and sex are two sides of the same coin which, if severed into two slices, is not a whole coin anymore" (1995, 83). He believed that the feminism that rejected males as the sexual enemy, and rejected heterosexuality as male-determined, was doomed to do no more than cause anguish, especially for feminists:

> Thus, in radical feminist ideology, it appears feasible to attack sexuality which is allegedly biologically determined and therefore immutable, without implicating gender which is allegedly socially constructed and mutable. According to this view, neutered gender, not sex, is the basis of male/female equality. The disparate male and female organs of procreation and their function are disregarded either as irrelevant, or as if they should be unisex, which is how radical feminism demands that gender should be. (1995, 82)

Money the feminist saw sexual liberation as the effect of birth control freeing women from the chains of reproduction, but others saw sexual liberation as just one more way of making sure women remained pornographic objects rather than world-changing subjects. Yet Money recognized something that continues today in issues such as the way women in the workforce are placed in the "mummy track" versus the "fast track." For whatever reason, more than forty years after the rise of women's liberation, society has been unable to accommodate female reproduction. Many would say that this just reflects continued discrimination on the basis of gender, but Money could make a strong argument that we have been unable to create an equality that can recognize "the disparate male and female organs of procreation."

In a comment on Kinsey, published in 1956, Money wrote,

> the scientific investigator who presents something new and unendorsed in the way of procedure, information or theory is always likely to be either worshipped or despised as a prophet. His worshippers will swallow whole his theory, method, and findings, their criticism submerged in adulation. Opponents of his work, too new or too radical for them to endorse, will brand it as anathema. (1956b, 233)

Money probably was aware that this projected his own future, although he seldom seemed troubled when he was submerged in adulation. In any case, the "new and unendorsed" became his banner throughout his career, and the more the new represented sexual freedom the better.

In 1961, he wrote an article for *Sexology* on premature puberty that seems to be a perfect example of the practical, common-sense Money:

> The chief goal of training with regard to masturbation is to have the child make it one of the private things of life never to be done in public. He need not be ashamed of doing it, but he certainly can be realistic about not bringing social disapproval upon himself by touching, rubbing or wiggling or squirming on his sexual organs. The child who cannot be realistically self-controlled in this way by the age of five is a special case almost certainly in need of seeing a psychiatrist. (1961d, 252)

One might say the same of his comment on swinging for *Gallery:* "Swinging institutionalizes honesty about the variety of sexual partners, instead of making it something that has to be done behind the scenes" (1977k, 58). Or on sexual television for Baltimore's alternative magazine *Harry* in 1991: "For instance, if I could have a television channel that was dedicated exclusively to sexual and erotic topics including basic information as well as entertainment, and I could have it nationally advertised that this was a very good channel to sit and masturbate to – when you're a teenager" (1991b, 7). He wrote a piece for *Sexualmedizin* in 1977 on sexual care for the handicapped: "Today, if a nurse/masseur were discovered masturbating an immobilized patient, he/she would be discharged, whereas to be decorated for moral courage would be more appropriate" (1977l, 962). One of his pieces for *Sexology,* however, apparently written in 1980, was not published:

> Blackmail, reprisal, exposure, separation, and punishment – these harbor more of the pathogenesis of incest than does the eroticism of sexuality with a member of one's own family ... Archaically, we still view incest not as the violation of a male by a female, but as the violation of female's virginity by a man who failed to pay a bride price for the right of possessing her. (1980g)

Perhaps even for *Sexology,* a pro-incest article went a bit too far. The second edition of Janice Raymond's *The Transsexual Empire* attributed the closing of the gender reassignment clinic to Johns Hopkins's irritation at Money's various comments that she typified as "pro-incest theorizing" (1994, xii).

Money always called for nonjudgmental objectivity in sexual science, but it is rather difficult to see this in his own work. He was nonjudgmental of sexual deviants and was quite ready to present any factual information, no matter how inflammatory, if it supported liberation. However, he was highly judgmental of anyone who took a conservative stance – often using pejorative adjectives – and he seemed completely unable to accept as objective any information that led to a conservative opinion. His support for male sexuality might seem conservative to some feminists, but it never led to any opinion that might be called patriarchal or sexist. In this sense, he was profoundly different from figures such as Hefner and Guccione. Instead, Money looked at the whole process in quite simple, logical, common-sense terms: if there were to be true sexual liberation, the majority of it would be likely to be heterosexual, given the dominant sexual object choices of both men and women. If it were to be heterosexual, then male sexuality would be essential. While Money did not at all countenance male aggression in the form of violence, sexual or otherwise, he saw the male pursuit of the female object as either innate or, more likely, so ingrained by history as to be unlikely to change in the near future. For female sexuality to be liberated, this truth of male sexuality must be accepted and somehow accommodated. This idea is so consistent in Money's writing that it is very difficult to accept that Colapinto believed in his depiction of Money in *As Nature Made Him* as the warped hater of male sexuality. Rather, Money hoped to allow all men and women to pursue whatever wild ecstasy they might desire.

CHAPTER EIGHT

As Nature Made Him

The Reimer Case

I mention the outline of the David Reimer story in the introduction to this book. One of identical twins had his penis burned off in a circumcision accident and, as a result, was brought up as a girl. Even that brief sentence suggests the potential impact of such a case. In the late twentieth century, gender became arguably the most charged aspect of Western culture: who is a man, who is a woman, and why? The penis was a central part of this discussion. To the popular imagination, it would be impossible for a case such as Reimer's to be other than significant. As depicted by Money, it changed the basic understanding of sexology. As depicted by John Colapinto in *As Nature Made Him: The Boy Who Was Raised as a Girl* ([2000] 2006), it revealed how much wrong can be done by a sexologist.

The story could be of the medical error in the burning of the penis in a Winnipeg hospital. The story could be of the many difficulties Reimer faced in childhood. We as readers might ponder the details about the family or the medical care: what might have made this story one of triumph rather than tragedy? But ultimately the story has become controlled by Colapinto. One colleague suggested that I am giving *As Nature Made Him* too much attention, as it is not a scholarly book. My first response is that it has been used as a source by eminent scholars such as Judith Butler (2004) and Anne Fausto-Sterling (2012) and that it is cited in hundreds of medical, psychological, and biological journals and books. This book has a scholarly currency much greater than the best article in the *Journal of Urology.*

My second response is that most of these citations use the book as a historical source, with little or no concern for the writing. As someone who has spent forty years teaching English literature, I always look at how the narrative works. Throughout this study, I have paid attention to what Money does with language, in particular to what his rhetorical flourishes reveal about his theories. I believe it is even more important to apply this method to *As Nature Made Him,* a book highly dependent on its heroes and villains. Just as any basic English course will tell you, the classic tragedy requires that the mighty be brought low. Colapinto depicts David Reimer not as the hero of "a tragedy" but as the central figure in a story of pathos. If there is a hero of a classical tragedy lurking here, it is John Money, in an albeit unusually unsympathetic tragedy. That erasure of sympathy has had a very strong effect on the reception of Money's theories in those many scholarly citations. Before Colapinto's book, Money's theories were often questioned. After Colapinto's book, they are usually damned.

Much of this chapter is about the way Colapinto presents his material. While I often became infuriated while reading *As Nature Made Him,* I also recognized that what Colapinto has accomplished is far beyond my resources. The cover blurb, from a review by *New York Times* journalist Natalie Angier, refers to the book as "brilliantly researched." Colapinto interviewed many who are now dead. Access to a lot of the material Colapinto read died with them. He dedicated himself to this story in a way that is just not possible for me as someone who is trying to offer an overview of Money's career rather than examining just this one case. While very few of the facts in the book should be disputed, the sum that Colapinto computes from these facts needs to be questioned. The back cover asserts that the story is of "amazing survival in the face of terrible odds." But it also says that it is "a macabre tale of medical arrogance." For Colapinto, this arrogance is not "medical" but personal – the arrogance of John Money.

Colapinto records a number of instances soon after the accident when the possibility was mooted of raising the baby boy without a penis as a girl, but he suggests this idea was only "a seed that had been planted" ([2000] 2006, 19). This seed "burst into full flower" when, on "their small black-and-white TV," the parents saw a transsexual accompanied by a psychologist, "a suavely charismatic individual in his late forties, bespectacled and with the long, elegantly cut features of a matinee idol." "Dr. Money's words, tinged with a highly cultured British-sounding accent, issued forth with uncanny fluency" (19). So they sought him out.

Angier states that the book is "cleanly written," which suggests, at least to this reader, a lack of metaphor, an absence of stylized enhancement. That "flower" jumping out of the tiny television in response to the "matinee idol" with the "uncanny fluency" is far from "cleanly written." Instead, it is part of a story with the highly controlled narrative found in certain kinds of detective fiction. The flower burst because it had no choice in the presence of this British-sounding accent.

I am not trying here to rehearse the facts of the Reimer case as recorded by Colapinto. Money and his supporters questioned certain aspects of the book, but, with the exception of those I explore below, most of these disagreements were quibbles about details (Heidenry 2000; King 2000a). However, the interpretive tone that Colapinto employs goes well beyond details and becomes the major factor in Money's present reputation. The furor about Reimer provided a very sad end to Money's career. At least to date, it has defined Money. My colleagues in gender studies tend to refer to Money as an egregious example of scientific error. Many Internet sites maintained by transsexuals and intersex individuals vilify Money. The majority of his obituaries devoted most of the discussion to the Reimer case, even those by sexologists who remained ardent supporters of Money, who found it necessary to spend less space praising him than to refute portraits of him as a monster.

This image of Money as monster has been created by *As Nature Made Him*. The book is the pathetic tale of David Reimer with a subtext of a Shakespearean tragedy of the King of Sexology. That subtext is not of Hamlet, however, a troubled figure who tries to do the right thing, but rather of Richard III, in this instance a monarch crippled not in body but in mind. The effect on Money's reputation is not unlike Shakespeare's effect on the popular view of the historical Richard III. *As Nature Made Him* was a "bestseller," but the breadth of its impact went well beyond that. There were a number of interviews on National Public Radio in the United States as well as television appearances both national and local, for both Colapinto and Reimer. I have found approximately one hundred North American newspaper reviews, some quite extensive and almost all very positive about the book and very negative about Money. Many people who know nothing about sexology know about the Reimer case. One of my graduate students mentioned my project to her mother, who replied, "That's that awful man who tried to turn boys into girls." *As Nature Made Him* has a presence that no academic book in sexology, even *Man and Woman, Boy and Girl* (1972d), could have – including the one you are now holding.

Money's views of sex and gender were always controversial. His earliest published work questioned the assumption that biology offers the definitive answer as to sex and gender. Money suggested in many publications that identity is a complex result of interacting forces. John Bancroft (1991, 2), in "John Money: Some Comments on His Early Work," refers to "the interactional mode of explanation which has characterised his thinking." That term, *interaction,* is central. It attempts to recognize all the elements involved in shaping the human. Genetics, or nature, is important, as is nurture, or the social environment, but so are other issues that are not easily placed in either category. "Nature" might be considered to be the condition before birth, but many environmental factors that enter the uterus shape the fetus. Prenatal and postnatal hormones are actually nurture, highly influenced by factors outside the individual's body, but often are considered as nature. Then there is timing: the stage of fetal development at which something external is introduced can completely change the effect of any aspect of "nurture," as has been clear in many medical problems, such as the thalidomide scandal. Perhaps the only pure "nature" should rather be the condition before conception. This creates an even bigger problem, however, as, on the one hand, nurture has a power over both egg and sperm and, on the other, there is actually no "nature" of the human before egg and sperm come together. As Money asserted in many different ways, in many different publications, the logic of nature versus nurture inevitably fails.

From the beginning, many disagreed with Money and instead asserted that biology provides the default order of human sex. One who presented this position strenuously was Milton Diamond, then a graduate student in biology. He believed that you are as your chromosomes define you. Some of Money's theories that seemed most inclined to promote nurture as the primary factor in gender identity were particular points of attack for Diamond. In the first article in which he confronted Money, in 1965, he suggested that Money and the Hampsons promoted a theory "that a sex role is exclusively or even mainly a very elaborate culturally fostered deception" (150). Diamond presented a clear challenge: "To support the theory of psychosexual neutrality at birth we have been presented with no instance of a normal individual appearing as an unequivocal male and being reared successfully as a female or vice versa" (158). One troubling element of Diamond's 1965 article is that he used animal studies to justify his assertions about human gender identity and then denied the arguments of the Hampsons and Money on the grounds that they used animal studies and assumed they are applicable to humans. This argument was central to his dismissal of imprinting, posited very early by the Hampsons and Money and continuing through much of Money's work. Diamond argued, "The basic incompatibilities are that learning requires reinforcement while imprinting does not and that imprinting is fixed and irreversible while learning is not" (165). Diamond arrived at the very open-ended conclusion that "the human being is extremely flexible" (169). Money's reaction was suggested by his 1971 review of *The Intersexual Disorders,* by Christopher Dewhurst and Ronald Gordon, which included a swipe at an unnamed person who would seem to be Milton Diamond: "The authors make the unfortunate mistake of comparing the views of experienced researchers and clinicians in the psychology of human hermaphroditism with those of an inexperienced student of anatomy who, on the basis of his experimental work with hermaphroditic rodents, published a rather ill-considered critique concerning sexual identity in human beings" (1971d, 217).

There is a large difference between the nature theories of Diamond and the interactionist theories of Money, but the difference is almost completely reversed when one considers the suggested practice. Diamond seems to believe that every infant should be assigned a gender based on chromosomes but that every adult should be given freedom to decide on gender, including whatever surgical intervention is desired. He presents the latter as a theory based on genetics in that someday someone will discover biological explanations of neurological differences in transsexuals and intersex persons that will offer a cause for why they are not normal (2000a, 2011, 2012). Thus, the transsexual desire for gender reassignment is justified by biology, but it is a biology yet to be determined.

It becomes a fascinating dichotomy. The Hampsons and Money assumed that intersex persons are normal neurologically but different anatomically. Their adaptability in terms of gender assignment, therefore, could be a model for humans in general. Diamond assumes they are abnormal biologically and thus probably abnormal neurologically. In this case, intersex persons offer no paradigm for those who are normal. For Diamond, some undiscovered biological cause justifies surgical intervention for transsexuals, as they need to rework their anatomies in accord with their abnormal brains. Another undiscovered neurological cause should prevent surgical intervention for infants who are intersex as they are abnormal and must await adulthood to find out how they as individuals wish to deal with abnormality. Diamond does not say this, but presumably, if the neurological abnormality could be discovered early, that abnormality will decide their adult gender identity, and then the confusion over their genders as newborns could be erased. For Money, surgical intervention could be justified for transsexuals as they wish to be the normal people they believe themselves to be, in the opposite gender. To be normal required appropriate genitalia – the justification to reassign David's gender as a female, to turn him from "abnormal" to "normal." In Money's view, the process simply conformed with surgical intervention for intersex infants, which could be justified because they are normal and require appropriate genitalia in order to function as the normal humans they are.

Diamond kept up the disagreement in typical scientific fashion through a series of articles. Still, there is no question that Money's theories were holding the day, receiving much more prominence than the biological alternative. A photograph of a professorial Money explaining a diagram of genitalia in a 1973 *Time* magazine article suggests the level of fame Money's ideas had attained. It was clear that the case of the boy who became a girl and the book *Man and Woman, Boy and Girl* (1972d) had changed the scientific view of sex difference and that this change was a major support for women's liberation.

The role of feminism in this process cannot be overstated. Money certainly saw himself as an activist in the drive to liberate women, partly because he saw it as an essential part of human liberation. In 1973, he wrote,

> Change in the cultural tradition of rearing males and females so that both may be liberated is for many people the opening of Pandora's Box – tampering with Nature, and letting loose options of Nurture that should be kept contained. Liberation is not tampering, however. It is simply the realignment of options that have always been available, though seldom utilized. (1973d, 4)

As all of us who engaged in feminist activism in the early 1970s can remember, the idea that gender difference was a social construction, and therefore social change could overcome the inequalities created by gender difference, was central. In the same article, Money said,

> To a degree greater than most people realize, gender-specific behavior is a matter of cultural heritage and is optional. The imperatives of sex differences may be summed up aphoristically in the saying that women menstruate, gestate and lactate; and men impregnate. (1973d, 5)

Since the 1970s, this assumption of the power of social change has been constantly thwarted, by everything from claims of absolute proof of differences between male and female neurophysiology to our inability to negotiate female reproduction as part of equality. "Gestate and lactate" could not be as simply accommodated as we might have thought. But as so often in early and mid-career, Money's theories were surfing just ahead of the zeitgeist and seemed to be perfect prophecies of the future.

Consciously or unconsciously, in the many books and articles in which Money referred to the Reimer case, such as *Sexual Signatures* (1975c), he maintained a tone that seemed to reinforce this sense of being within a general theoretical movement. Colapinto ([2000] 2006, 104) suggests that *Sexual Signatures* "was couched in the language of commercial poppsych bestsellers," which is quite inaccurate, but the book is very interesting for another reason. In most of his books, Money tried for a tone of scientific objectivity, but his tendency to use inflated language produced an emphatically personal style, and the necessary scholarly apparatus produced a constant reiteration of citations of his own name in parentheses. This could create an impression that the arguments were somewhat centripetal, Money talking about Money. This is much less evident in representations of the Reimer case. *Sexual Signatures,* co-written with Patricia Tucker, a journalist, almost erased Money from the process. The account in *Sexual Signatures* of the moment when the Reimers saw him on television referred only to the transsexual, not the unnamed psychologist with her, and the narrative was well along before "the medical psychologist described the alternatives to the parents" (1975c, 93). The result seems to be a dispassionate representation of a simple reality: this is the way things now will be done. Thus, the famous man in *Time* was also the recorder of an inevitable process.

Such notoriety just made Diamond's concerns deepen, as this one case seemed to be the linchpin of Money's theories. Diamond decided to seek greater information. According to Colapinto ([2000] 2006, 199), Diamond placed an advertisement a number of times in the 1980s in "an American Psychiatric Society newsletter" that said, "Will whoever is treating the twins please report." A psychiatrist who had treated Reimer, Keith Sigmundson, eventually replied. Money saw as a significant example of Colapinto's sloppiness that the society had only the *American Journal of Psychiatry,* which never carried advertisements. Regardless of the venue, Diamond reached Sigmundson, and the result was eventually their 1997 article "Sex Reassignment at Birth: A Long Term Review and Clinical Implications." The obvious news value of this revelation almost immediately led to Colapinto's (1997) *Rolling Stone* article, "The True Story of John/Joan," which ran as a typical pop culture exposé of the case, followed by the book.

There can be no doubt about the error in the treatment of David Reimer. Colapinto and Diamond assert simply that Money's theory could not work, that a boy could not be turned into a girl, but this is not the only explanation of the error. Another possibility is that David Reimer as an individual would not be transformed from boy to girl. This is essentially the position of both intersex and transsexual activists: a person was transformed from one sex to another in order to give him a gender identity other than the default and other than his choice. For them, if a person accepts the default, all right, and if not, all right, but it is the person's choice, not to be made by doctors or parents. The social or psychological reasons why David could not become a girl are irrelevant: the point is that he did not want to be a girl. A third possibility, however, noted but not accepted by Colapinto, is that the transformation was made too late. In 1955, Money had written, "Though gender imprinting begins by the first birthday, the critical period is reached by about the age of eighteen months. By the age of two and one-half years, gender role is already well established" (1955c, 310). By the time Money was contacted, Reimer was nineteen months old, and the first feminizing surgery was performed three months later. If Money's "critical period" were correct, then Reimer had already entered a questionable zone.

Regardless of the reason the error proved to be an error, Reimer's female gender could have been reversed. Money's studies of intersex individuals mention many who decided to change their gender assignments later in life. Thus, Reimer's choice to return to being a male was unusual but no more than that. One person's reassignment need not have been extremely troubling to Money's theories if not for the unique position it took in the history of sexology. One can easily see why Money might have been more committed to this gender assignment than others: much depended on Reimer remaining female. It is possible – and it remains only a possibility – that Money's arrogance and obsessive self-promotion led to the gender assignment continuing long after it should have been changed. Money also, however, might have had hopes that the apparent "maleness" that Reimer demonstrated as a child, which Colapinto and others suggest shows the error of gender assignment, could have had a successful reconciliation with a female gender identity. In an article from 1977, Money called for "Destereotyping Sex Roles:" "Thus a tomboyish girl, prenatally androgenized, grows up to be a career-minded woman, not a transsexual who claims to need sex reassignment" (1977d, 28). Money would have believed that a constant reinforcement of Reimer's female identity by all around him, including affirmation of feminine gender roles that Money himself would probably have found stereotypical, would ultimately result in a female identity, albeit one that would have been unlikely to be very feminine.

In 1998, Money published "Case Consultation: Ablatio Penis," the medical term for detachment of the penis. It apparently was completed in 1996 and thus predated both the article by Diamond and Sigmundson and the one by Colapinto. Money stated,

> The idea of sex reassignment as a sequel to infantile genital trauma or mutilation is, in history, a phenomenon of twentieth century medicine. Its historical recency no doubt accounts for a good part of the unresolved controversy which it still engenders and which perpetuates the absence of the very data upon which the scientific resolution of that controversy is contingent. (1998a, 122)

Yet, while Money seemed inclined towards sex reassignment, he stated that there had been too few cases of ablatio penis to make an informed decision as to which choice should be the standard of care: "too few cases are concentrated in any one treatment center for a statistical, matched pair study of the variables that affect the outcome of treatment with and without sex reassignment" (1998a, 121). In other words, Money was wrong to say that the success of the Reimer case showed the efficacy of sex reassignment in ablatio penis, and Diamond was wrong to say that the failure of the Reimer case showed that it didn't work with ablatio penis. The best argument against Diamond is probably the case reported in Bradley et al. (1998). The article is titled "Experiment of Nurture," perhaps in argument with Colapinto's then very recent *Rolling Stone* article. The Bradley (1998, 4) article presents exactly the same narrative, a boy with a penis burned off in circumcision, but a contradictory result as he became a woman who identified as a female and claimed to have no doubt of her gender identity: "The most plausible explanation of our patient's differentiation of a female gender identity is that sex of rearing as a female, beginning at around age 7 months, overrode any putative influences of a normal prenatal masculine sexual biology." This would suggest that Reimer simply was treated at too late an age.

There are various aspects of the substance of Colapinto's description that seem strange. For example, he states that in the school that Reimer attended as a child "one girl was an intersex" ([2000] 2006, 132). Almost no intersex persons are identifiable by appearance when clothed, and it is very unlikely that any child in that period would so self-identify. So on what basis is this claim made? Similarly, the Reimers claimed that Money introduced the twins to a transsexual who tried to influence them. Throughout the narrative, Colapinto refers to this person as "the transsexual," as though this is an appropriate way of identifying someone. Change the words to "the homosexual" and the pejorative import becomes obvious to anyone. In any case, Gregory Lehne says that the event did not take place but that there was a born woman who worked with Money at the time who had a decidedly masculine demeanour. That a very distraught David Reimer should recall her as a transsexual is not surprising. Both of these examples create an atmosphere of gender confusion, quite appropriate to Reimer's position but perhaps not the objective factual evidence that Colapinto claims.

Many aspects of Colapinto's representation of Money himself simply do not conform to the impression conveyed in the conversations I have had with his colleagues and patients. Part of the explanation could be because the story the Reimer family told him created a very different portrait. Colapinto recounts that the Reimer brothers claimed that Money tried to make them simulate sex with each other. No one else has suggested such behaviour, in spite of Money's constant assertions of the benefits of sexual rehearsal play among children. He believed it to be a useful part of sexual development but did not make it part of his clinical practice. Instead, this account contributes to Colapinto's overall portrait of Money as a sex-mad pervert. Colapinto ([2000] 2006, 98) says that, in *Sexual Signatures,* Money "even went so far as to recommend that parents engage in sexual intercourse in front of their children." On the contrary, *Sexual Signatures* quite correctly observes that in many cultures parents sleep in the same room or even the same bed with their children and still have intercourse. Money's suggestion was not that American parents should put on sex shows for their children but rather that, if American children chance to observe sex, the parents should not panic in revulsion but rather make it an educational moment (1975c, 135).

Another surprising part of the book is the claim that, when the Reimer parents were present, Money was considerate, gentle, and responsive, but when they were absent he treated the twins in a belligerent and even oppressive manner. This is quite contrary to many famous stories about Money that contrast his belligerent and oppressive treatment of his colleagues and underlings with his fatherly, supportive treatment of patients. The patient I have called "Bob" told me that it was not uncommon for him to be in a private conversation with Money when some colleague came by. At the same time as he sweetly listened to him, Money would let loose some foul-mouthed invective at the inadequacy of the colleague. As Bob told me, "It was extremely embarrassing."

Part of the basic narrative of the book is that "Mickey Diamond" slew the dragon-like John Money or, as Money himself put it, "Plucky little David from faraway Hawaii confronted the big Goliath of the medical establishment, The Johns Hopkins Hospital, in Baltimore" (1998b, 321). No story can be without its colour, of course, but Colapinto ([2000] 2006, 40) takes this to extremes: "The son of struggling Ukrainian Jewish immigrant parents, Milton Diamond, whom friends called Mickey, was raised in the Bronx, where he had sidestepped membership in the local street gangs for the life of a scholar." The depiction of Money is much more detailed, but that expansion is about his psychologically damaged family and makes no mention of the economic strictures of his childhood. The struggle of Money's family was such that he spent his life obsessed with poverty. His family may have been damaged, but there is no reason to suggest that Money was some severely warped individual. Colapinto misquotes Money's own words to turn him into a self-castrating man hater, as noted in Chapter 1 of this book. This is interesting given that one consistency in all recollections of Money is a decidedly masculine form of arrogance. Colapinto's misquotation allows Norman Doidge (2000), in a review of *As Nature Made Him,* to use this passage to prove that the metaphorically castrating Money should be connected to Alfred Kinsey's self-circumcision. Doidge's repetition of the misquotation implies that that is what sexologists are like.

Colapinto ([2000] 2006, 277) states, "while I consider David's case to be among the strongest evidence yet available for the biological underpinnings of gender, I reject any reading of the book that reduces his story to simple-minded biological determinism." Diamond has said much the same. However, it is difficult to see their arguments as much more than biological determinism. Colapinto's title is taken from Rousseau, but the source is ultimately window dressing. The book asserts what the title suggests, that Reimer was nothing except what nature ordained. The tone of the book tempts me to see the source as more than nature, "As God Made Him." I know nothing about Colapinto's religious beliefs, but he seems to suggest, to mix two biblical references, "what God has wrought let no man put asunder." Money himself saw the eager embrace of Diamond's discoveries by so many as conservative, religious, and anti-feminist. In a 1998 letter about an article in the *New York Times,* he said the whole purpose of the antagonism towards him was to assert "proof that biology is destiny, and that both feminism and all of social science are in error" (in JMCH). It is noteworthy that Money's only significant published confrontation with Colapinto's view is in a book titled *Sin, Science, and the Sex Police* (1998b).

In Colapinto's book, the rejection of Money's ideas is often superseded by a rejection of Money the man, a man whom he depicts as all-powerful. Colapinto writes as though Money made all the decisions, while anyone familiar with medical procedures and particularly surgery knows how little power any psychologist can have. Some of Money's colleagues recall him almost pleading with urologists for the opportunity to do psychosocial follow-up with their patients. While surgeons had a number of ways of ensuring that postoperative patients would see them, Money had nothing except the generosity of the patients and their parents and his own ability to appeal to their commitment to their own education and to that of others.

One of the damning indictments of Money is that he never recanted. Even if he continued to believe his treatment regime was correct, which he did, he should have admitted that his central example had not proved successful. His excuse was always that Reimer was "lost to follow-up" because he refused to see Money anymore. Money had a reasonable argument in that a patient who does not wish to continue as part of a study is "lost to follow-up," even if he or she lives next door. Still, in this case, "lost to follow-up" seems just too self-serving, as does his rejection of the filmmakers who eventually found Reimer. Money asserted that the British journalists who searched out Reimer showed no care for the privacy of the patient. Their tactics were certainly questionable. But Money at times offered case studies that in theory were not identified but were sufficiently revealing that anyone could find the subject with very little difficulty. Then again, Colapinto's justification of the British tactics on the ground that the world needed to know seems similarly specious. Money used ethics to hide the results, and Colapinto explained away unethical behaviour that helped him to confront Money's ethics.

Ultimately, these are minor issues in the overall narrative. There seems to be no question that Money was the guiding force at least postoperatively. Before the operation, it was his presentation that convinced the parents that this course of treatment would work. And there is no doubt that he was ultimately wrong and that he did very little to let the world know that his course of treatment in this instance had failed. However, Diamond and Colapinto see this not as a course of treatment but rather as an "experiment" that destroyed a human life. Money was committed to research to an extraordinary degree, even for a high-achieving medical psychologist. Just as one example, there is a study of whether males born with a micropenis have any unusual features in

IQ (1983f). The study notes that the cases examined show a higher than average IQ but that this is unlikely overall to be attributable to the sex ambiguity. As a non-scientist, I find such a study to be almost absurd: did anyone suggest that these children did not have normal IQs? But Money's voluminous publications suggest that he was willing to study almost anything and publish it. There seemed little doubt that he yearned to "experiment" in various ways.

In a case such as this, the word is damning by itself. We associate "experiments" on humans with Josef Mengele and the Nazis. The fact that Reimer was an identical twin makes it even worse. Anyone with even the slightest experience of science knows how important twin studies have been as a control on almost any process. Mengele's twin studies are among the most infamous of Nazi experiments. The justification for use of the word is Money's own comments, which appear so often in his writings as to become liturgical. This is one from *Man and Woman, Boy and Girl:*

> In the case of human development, the glimpse cannot be created at will, as in planned breeding experiments, because of the ethical limitations that we human beings place on our unwarranted interference with one another's lives. One must rely instead on Nature's own experiments – the quirks and errors of chromosome sorting that occur spontaneously. In the case of animals, planned experiments can be designed. (1972d, 25)

For Diamond and Colapinto, this is proof that Money was using Reimer as an animal.

The obvious retort is that Money clearly thought of this as efficacious treatment. While he mentioned the twin often as part of his explanation of the case, he was also aware that the existence of a twin could cause problems. He stated that, given the obvious presence of Reimer's brother, the medical history would need to be revealed to Reimer at quite an early stage, perhaps earlier than was advisable. In other words, the twin factor was by no means wholly advantageous. There is no reason to assume that he was more eager to apply this treatment because Reimer was a twin. His many articles on the treatment of boys with a micropenis show his belief that any boy without a penis viable for heterosexual intercourse should be brought up as a girl. One of his micropenis patients, who now seems to be happy as a gay man, told me that he believes he would have been better off had he been given a female gender assignment, but, by the time he was brought to Money's clinic, he felt his life was set. In 1977, Money published a study called "The Successful Use of a Prosthetic Phallus in a 9-Year-Old Boy," but he clearly thought the "success" was limited, and he described the problems of some twenty patients with a micropenis: "In the best of all possible worlds all of the professionals, from the delivery room onward, associated with the care of a baby born with a micropenis would know how to talk to the parents in terms of having a daughter whose genital anatomy will need correction" (1977i, 194). Richard Green told me in conversation that he thinks Money gave the right advice to the Reimers, given the standards of care at the time, although Green would not agree with this approach today. As with any medical procedure that is later superseded, hindsight cannot deny this.

Of course, Green is just one of those Colapinto ([2000] 2006, 249) describes as "Money's acolytes." Colapinto's tendency to spin any observation to attack Money and support his opponents is decidedly irritating for any reader who might wish for a dispassionate account of the history. Anything to do with Money is negative, and anything to do with Diamond and Reimer is positive. Just as one example, I offer the following. This is a lengthy description of Reimer's wife, but a careful examination of the language, particularly the adjectives, gives a sense of how Colapinto's writing shapes his story:

> Twenty-five years old, Jane Fontane was a pretty woman with blue eyes and shoulder-length strawberry blond hair. At five feet one and one hundred and eighty pounds, she was sensitive about her weight, but she carried her generous size easily, and to those who knew her, it seemed merely a natural adjunct to her nurturing personality. When I first met Jane in the summer of 1997, her combination of unflappability, affectionate friendliness, and infectious laughter reminded me of no one so much as the central character in Joyce Cary's comic novel *Herself Surprised* – the unsinkable Sara Monday, the picaresque mother of five children, a woman whom Cary describes as a kind of force of nature, a woman whose earthy goodness and fundamental optimism see her through every scrape life can throw at her – including her own youthful poor judgments.
>
> Like Sara, Jane possessed a guilelessness and innocence that helped to explain how, by the time she met David, she was herself the single mother of three children – by three different fathers. Unworldly to a fault, Jane was a lifelong nonsmoker and nondrinker, a homebody who did not go to bars and didn't approve of "cursing."

> Her chief flaw was a certain neediness, a result perhaps of her difficult childhood in Winnipeg, where she was raised by her mother and stepfather. (192)

I know nothing about Jane Fontane except what I have read. Colapinto's description of her, however, seems to me an extraordinary example of how a writer can take details that could easily be represented as negative, a very overweight twenty-five-year-old with three children by three different fathers, and instead turn the person into a paragon of the female as she should be. The negatives are dismissed as "youthful poor judgments" and a "certain neediness," far surpassed by her "earthy goodness," which makes her a "natural," "nurturing" "force of nature." She is at once the equivalent of a beloved character of fiction, with all the vital resonance that Joyce Cary could provide, and also, quite clearly, as female as it could be possible to be – as nature made her.

The contrast with the portrayal of John Money is severe. Money is represented as a patrician person of privilege who wields power like a medical warrior. His pleas for sexual liberation are at best the decadent reflection of not a "difficult childhood" but rather a warped, violent upbringing in a weird land of perverse masculinity called New Zealand. When that liberation included children, he simply went beyond the pale, as in his supportive foreword to *Boys on Their Contacts with Men* (Sandfort 1987), which Colapinto describes with dripping distaste as "an unusual volume" ([2000] 2006, 30). Throughout his book, Colapinto attacks Money on issues not related to the Reimer case. Instead, the observations suggest that Colapinto wishes to increase the American public's association of Money with evil by demonstrating his support of sexual activities that the majority would find unacceptable. At times, it looks like the Marquis de Sade became a medical psychologist at Johns Hopkins University.

That last comment is no doubt too extreme, yet a number of reviews note the possibility that Colapinto goes too far in this aspect of his book – that Money is attacked too strongly. Angier (2000, 10), one of the strongest supporters of Colapinto's book, states, "Money emerges as too evil to be believed." Chris Bull offers a similar but more extensive observation:

> The book falters only in its depiction of Money. Though the scientist is easy to dislike for his apparent arrogance and blinding ambition, the author, perhaps seeking the perfect antagonist for his narrative, reduces him to a Frankenstein who forces pornography on his unwitting patients and sympathizes with pedophiles. Though

> his handling of the twins case arguably amounts to malpractice, Money was clearly motivated at least in part by compassion for a patient whom other medical professionals had abandoned. Determined to make an argument against Money, the book overlooks the important contributions he has made in his career as a whole to our understanding of sex and sexuality. (2000, X09)

Colapinto's vision does posit a Dr. Frankenstein, someone who turns the innocent baby, already damaged by the errors of science, into a monster. Near the end of the book, the level of demonization goes beyond this when Colapinto presents an anecdote about an evening spent with the Reimers in their home. Both parents wanted Colapinto to watch their favourite movie, *Crossroads*. As Colapinto notes, the film fictionalizes the famous story that the blues legend Robert Johnson made a deal with the devil to become a great guitar player. Of course, the Faustian pact was a common legend long before this, long before even the various literary versions of Faust, and *Crossroads* is not one of the better examples. The Reimers suggested that their primary attraction was because they love the blues. And the blues soundtrack by Ry Cooder and Steve Vai is undoubtedly the best feature of the film. However, Colapinto sees much more in the scene in which Blind Willie Brown laments that the devil did not give him what he wanted:

> Both Ron and Janet hung on every word of this dialogue – as if they expected that on this viewing, the scene might finally play out differently. When the scene ended, Ron sat back in his chair, then glanced quickly at me and away. Several times during our long interview that day, I had tried to get Ron to speak about how he now felt about John Money and the momentous decision he had convinced Ron and Janet to make. He had made a few halting, stumbling efforts to answer my question but had clearly failed to say all that was in him. Now I felt I had my answer. Along with Ron's grief and guilt there was an obvious admixture of outraged betrayal, which lay too deep for him to express in words. ([2000] 2006, 256)

The Reimers had met Money at the gender crossroads and had sold their souls.

This is no doubt the most extreme example of the failure of Colapinto's claim to "scrupulous objectivity" ([2000] 2006, 245), but there are others. One of particular concern given the way he portrays Money is the representation of "Paula." Born a male with ambiguous genitalia, Paula was brought up as a female. Colapinto begins the narrative, "According to the *Time* magazine article, the sex change had been a complete success" (226). The obvious implication is that it was not. According to his own description, Colapinto pressed Paula incessantly, to no avail: "I asked if, while growing up, she had ever thought, 'Maybe I'm a boy,' but again she spoke before I could get the question out. 'Never,' she said. 'Never at all'" (226). According to Colapinto, Paula's delusions are exceeded only by her mother's: "To this day she refers to Money as her 'savior' and speaks in only the most glowing terms about him and about her former son's conversion to girlhood" (229). Colapinto tellingly notes Paula's father's "severe clinical depression" (230) and Paula's refusal to deal with her past: "She has turned down a recent request from Johns Hopkins to participate in a follow-up study on sex reassigned patients. She simply does not want to relive her childhood – which she nevertheless insists was a perfectly happy one" (231). Colapinto seems to find it revealing that she gave up her early ambition to be a lawyer in order to become an obstetrics nurse: "Today she helps to deliver babies in the same small hospital where she was born twenty-seven years ago, as Michael Edward" (231).

Clearly, Colapinto believes all aspects of the Reimers' point of view of David's story. The family developed a narrative, and Colapinto extends it. The dreadful circumstances no doubt were as the Reimers described, but Colapinto's spin on the descriptions provides most of the flavour of the representation, a flavour that contributed significantly to Money's demonization in the popular press and even in the intersex and transgender communities. But why did Colapinto do this? Was it just the combination of being convinced that David was a victim and of being a journalist trying to create the best story? It is certainly a far better story than that offered by Money's supporters, the attempt by a knowledgeable and committed professional to help someone severely damaged by medical error, an attempt that proved a failure. But perhaps I can turn a Colapinto narrative back onto Colapinto.

Both Money and Colapinto regarded the death of Money's father when Money was eight as an important note in his development. Colapinto calls Money's father "a brutal man" ([2000] 2006, 26), and Money would have had little disagreement. In contrast, Colapinto recalls his own father with affection as a great storyteller, the person from whom he originally heard of the Reimer case. Colapinto's father was chief of urology at St. Michael's Hospital in Toronto: "He died in 1985 of hepatitis B contracted through a seemingly innocuous prick of his finger while operating on an infected patient" (3). "He was also the co-author of a book with the snappy title *Urological Radiology of the Adult Male Lower Urinary Tract*" (3). Colapinto's mother was a nurse who, late in life, published some short fiction.

So Colapinto was the child of a proper surgeon who wrote a medical treatise with a properly anatomical and boring title. His father died as a result of his sacrifice to that proper surgery. Unlike Paula, Colapinto's mother was born to be a nurse and pursued her avocation as a writer only after she had done her proper maternal and nursing duties. This is no doubt unfair, although I am following a narrative that Colapinto has already spun. Still, it is difficult not to see a contrast, perhaps even a self-conscious contrast, between Colapinto's representation of this medical family and the childless and only briefly married non-medical doctor Money. In the same way Colapinto uses positive and negative narratives and descriptions to create heroes and villains, one can interpret his representation of his own family as descriptions of heroes that create the writer that almost inevitably represented John Money as the devil.

One interesting aspect of the whole Reimer case was given rather little attention by Money, Diamond, or Colapinto – the fact that the Reimers were identical twins. For Money, this was a fortuitous factor in terms of the study, a likely source of difficulty in the treatment. To Diamond and Colapinto, this was part of the reason for the "experiment," the existence of a control of which the only difference would be gender. However, none of them seems to have given much attention to what might be called the magic of identical twins. In the "Tragic Update" published in the 2006 paperback edition of *As Nature Made Him,* Colapinto reflects on Reimer's suicide in 2004 but first mentions his brother's 2002 death from an overdose:

> Despite the intractable emotional impasse at the center of their relationship, the brothers did share the preternatural closeness often observed in identical twins. Both told me they could practically read each other's thoughts. (12)

The classic study of identical twins, now known as the Minnesota Twin Study (Farber 1981), suggested that genetic code explains why twins reared apart are often eerily similar, but this seems insufficient. Diamond himself has since used identical twins for a study of transgender (2013). The tabloid press depicts identical twins who grew up in different locations as two humans who are psychologically conjoined. There seems to be something much more in the connection between identical twins than the nature in which Diamond and Colapinto have so much faith, but also something much more than the nurture and hormones that Money hoped would produce a girl.

A number of Money's supporters have inferred that the Reimer family was probably a more significant factor than Colapinto suggests, given the mental disorders and substance abuse he documents in the family. *As Nature Made Him* represents most of the problems in the family as likely to have been caused by David's injury and treatment. Colapinto gives little attention to implications that all would not have been well if the twins had had no medical problems. Still, I would always return to those identical twins. I wonder whether science could ever understand what operates in identical twins. Regardless, perhaps all involved should have read a brief article published in 1960, five years before the Reimers were born, in that periodical that later became so key, the *American Journal of Psychiatry*. David Swanson's information about one more strange link between the monozygotes is simply summed up by his title, "Suicide in Identical Twins."

The importance of the Reimer case for Money cannot be overstated. From the publication of *Man and Woman, Boy and Girl* (1972d), it was a central example in many of his books and articles. He attached his theories and his career to this one case. Part of the effect was to set up a narrative that was precisely what Milton Diamond did not believe. When the narrative proved to be false, Money was doomed. All of his theories, many of which did not depend at all on the Reimer case, were doomed. Many would expect Money to have reacted more vociferously, in defence not only of himself but also of his work. Part of the reason he did not was Johns Hopkins, which did not want him to respond as the hospital saw that nothing was to be gained by fighting and that much could be lost. Many of Money's colleagues tried to support him. "Bob" attempted to establish a group of those who thought their lives had been changed positively by Money. John Heidenry wanted Money to write an extensive rebuttal, but Money refused. The result was Heidenry's own review of *As Nature Made Him* and Milton Diamond's reasoned response to Heidenry (Heidenry 2000; Diamond 2000b). Michael King wrote a brief, supportive profile piece, again with a response from Diamond (King 1998; Diamond 1998b) and a questioning review of *As Nature Made Him*, both published only in New Zealand (King 2000a).

Besides "Ablatio Penis" in *Sin, Science, and the Sex Police* (Money 1998b), a rather hidden few pages, Money wrote little about the case after Diamond's first revelations. His responses, primarily in personal correspondence, tended to be about details of facts, assertions that Diamond was always pursuing him, and claims about the unethical behaviour of others. Whether an admission of his own errors would have in any sense helped Money's long-term reputation is an open question. One telling bit of evidence as to Money's own feelings is in the Money collection in the Kinsey Institute. Money was always very proud of his fame and kept a bound collection of his clippings. The collection ends with the first mention in the mainstream press, in the December 1996 *Esquire,* that the famous boy who became a girl was not as she might have seemed (Segell 1996).

Conclusion

Venuses Penuses

Vern Bullough's brief article in praise of John Money included this paragraph:

> Money was a hard man to convince to change once he made up his mind. I tried to get him to change the title of *Venuses Penuses,* a book for which I was general editor, but he was determined to keep the title. The title of the book to my mind had little relation to the subject matter, which dealt with philosophy of sex and sex research, but my arguments were to no avail. It was John's book and not mine, and I eventually abandoned my arguments. Not infrequently in our relations, he went his way and I went mine. One is advised not to argue with John. (2003, 235)

Money's own justification of the title was given in the foreword to the volume, a three-verse limerick that Money stated he wrote in 1953, which he called "Venus's Penises." He concluded with this:

> The publisher and his advisers had some misgivings about being able to market the book under its poetic new title. A search of books in print revealed no listing of a book in the English language that included either the word *penis* or the word *vagina* in the title. Here was

> a challenge! It was surmounted by means of invoking an artifice of medieval Latin orthography, by which the title was transmogrified into *Venuses Penuses* (correctly pronounced as Venoos' es Penoos' es). (1986d, xx)

Of course, the title is not "poetic" but silly, a bit of schoolboy humour. As Bullough suggests, it is inappropriate for the book. But the book itself is very strange, even for a book by John Money. It is six hundred pages of his articles. Money stated,

> In 1982 I made a collation of journal articles from my own bibliography and had three sets of them bound and titled *Principia Theoretica*. This volume became the forerunner of the present publication. Each paper has been selected for inclusion on the criterion of its theoretical significance. (1986c, 3)

The choices of articles and the arrangement seem rather arbitrary. There are many that are redundant and do not seem unusually significant. That original title for the self-bound volumes suggests Money's own view of their significance, something to rival Newton, Descartes, or Russell. In other words, both title and collection clearly represent the bull-headed arrogance of John Money.

But they also represent the independence and imagination. I doubt that anyone except me will take the time today to read all of *Venuses Penuses*, but the excursion is more rewarding than a perusal of all of Money's other books. While it would have been a much better book if the different parts had been digested and reshaped, the number of strange sparks would have been diminished. Not only does this represent a fascinating mind at play, but also it shows most of the best ideas in sexology to come out of the late twentieth century. Money might not have been Bertrand Russell, but he is to date probably the only example of a Russell-like brilliance brought to bear on the problems of sexology. Never as systematic and coherent as Freud, he had an independence of thought unmatched by Kinsey or Masters. I titled this chapter "Venuses Penuses" rather than using some phrase more obviously appropriate to a conclusion because the title is the summary chosen by "The Man Who Invented Gender." As Bullough implies above, Money was often his own worst enemy, especially in his tendency to offend his greatest supporters. Perhaps the impossibility of summing up his successes and failures is represented in the impossibility of that title, "Venuses Penuses," something that should have been stopped before he even presented it to his editor. And yet the joke, the pun, the rhyme, the combination of the classical god and the medical anatomy, the schoolboy Latin at play, all suggest Money's obsessions with what language could do and how it could be used to expand and enhance his research and his influence on the world of sex and gender.

The title suggests why there was so much wrong with what he did, so much that covered up that brilliance and insight. Part of it was that independence, but an independence that often seemed without restriction. In the introduction to *Venuses Penuses,* he assessed himself:

> I maximize what I have come to consider my chief intellectual asset, namely, an ability to put conceptual order into informational chaos, and an ability to relate concepts to one another to form systems or hierarchies of order. This type of ability needs constant, empirical monitoring, for without it, I can run wild into paranoid, delusional systems. Before the empirical checking is done, I have found that a new hypothesis or theory can be somewhat frightening, for there is no guarantee that it will not be considered crazy by other people, especially if it is unorthodox and against the current climate of one's discipline. (1986d, 11)

Many have called Money's ideas "delusional" or "crazy," and they have certainly been unorthodox. Still, this spirited freedom from restraint was his source of intellectual power. However, the "empirical monitoring" was not as successful as it might have been. This is perhaps not surprising, given the chain of logic in the last sentence. According to this, the "new" that is "unorthodox" will be treated as "crazy" until the "checking is done." There is little suggestion here that the checking might contradict those theories and justify the assumption that they are crazy. I have read nothing in Money's work where he recognized one of his "systems" to be "paranoid delusional" and so rejected it. Few of those who worked with him recall him listening to them. All of the colleagues I have interviewed recall a man who easily became belligerent when crossed. In his brief autobiography for *The History of Clinical Psychology in Autobiography,* he concluded with an epilogue titled "The Dissident" in which he stated, "I have on several occasions been a sitting duck for the potshots of other professionals" (1991a, 269). But as so often seems to be the case, someone who consistently views himself as the persecuted truth bearer is remembered by others as simply unwilling to accept the possibility that what he was bearing was not the truth.

The publisher he chose for *Venuses Penuses* might be an example of Money's view of his theories. While his early work was published by Johns Hopkins and he had books published by Oxford, Little Brown, and a number of others, his later work was usually published by Prometheus Books. Founded in August 1969 by Paul Kurtz, who also founded the Council for Secular Humanism and co-founded the Committee for Skeptical Inquiry, its history reflects its name and its founder: arbitrary and fiercely independent. Prometheus decides what fire to bring where without any concern for what others might think, and Prometheus Books published what it wished to publish with little concern for sales or reputation, except to be controversial. In Prometheus, Money found a publisher that would let him create the books he wished to create. He could reject the advice of editors who sought consistency or coherence.

Or rigour. As editor of *Archives of Sexual Behavior*, Richard Green would be confronted by Money the assessor when he found other scholars to be less than precise but then confronted even more vigorously by Money the author whenever anyone questioned his own work. The bibliography to *Venuses Penuses* suggests part of the problem. This is the "Author's Explanatory Note":

> This bibliography of publications from the Psychohormonal Research Unit has a history of having been compiled annually. In 1974, the years from 1948 through 1974 were consolidated, with separate categories and numberings for books, scientific papers, reviews and book chapters, et cetera. The second consolidation covers the years 1975-1979. From 1980 to the present, there is a separate supplement annually. However, the numbering within each category of publication is continuous, not annual. All books are listed continuously from 1952 through 1986. On coauthored scientific papers, the name of the senior author is not invariably listed first, but sometimes last. The responsibility of the senior author, as director of the Psychohormonal Research Unit, was to design, supervise, write, and edit each paper. (1986d, 613)

A bibliographer who read that would not know whether to laugh or cry. If I were given that by a doctoral candidate, I would make it quite clear that it is not acceptable. Presumably, someone might have said to Money that he should rationalize this listing to make it coherent. And presumably, if someone said that, Money replied that it wasn't necessary. He never claimed to be a bibliographer, and many might say that a consistent bibliography is irrelevant to the work of someone such as Money. Yet, as any academic knows, it is through a readily accessible bibliography that a reader has access to the material to monitor the ideas represented. Without such resources, the reader is left little choice but to accept the theories as given.

Colapinto records a number of examples in which Money seems to have been less than scrupulous in his care for the privacy of his subjects. I have noted that early in his career Money used Janet Frame as an example of psychopathology in a paper that did little to hide her identity. This seems decidedly out of keeping with his own claims about the necessity of confidentiality in the Reimer case and others. A slightly different problem is visible in one of his examples in *Lovemaps,* in which he commented on Stockholm syndrome, where a captive falls in love with the captor:

> A well-published example from 1976 is that of the Baltimore pedophiliac lust murderer, Arthur Goode (Waters, 1984). At the age of twenty he abducted a newsboy from a professional family and coopted him as his boy lover. (1986c, 46)

Money presented the boy as a willing accomplice, a view that was questioned at the trial and elsewhere. The source Money listed is not a scholarly or legal one; rather, it is an interview with Goode made by the filmmaker and collector of sexual weirdness, John Waters. Waters quite clearly was fascinated with Goode, but the article represented Goode's view of the world rather than Waters's view of Goode. The article, published in Baltimore's *City Paper,* made no claim as to the veracity of any of Goode's narrative of events, much less of Goode's opinion of what the boy's feelings were. Yet Money seemed to have used this as his only source. Waters elsewhere referred to Money, admiringly, as the Duke of Dysfunction (King 1998). Perhaps as a duke Money felt the right to be rather cavalier about his sources.

While Money had an interest in sexual psychology from an early age, in the beginning of his career he often seemed more of an academic entrepreneur than a dedicated sexologist. The dyslexia clinic he planned was the most significant example of a tangential move away from sexology, but there were others, large and small. Thus, *A Standardized Road Map Test of Direction Sense,* first published in 1965, seems to have resulted from Money's recognition of spatial disorders in Turner syndrome. The first edition was a small production by Johns Hopkins (1965h); ten years later, however, a new edition was produced by Academic Therapy Publications, one of the largest companies in the field, whose material is found in almost any American psychotherapist's office (1976d). Money appears to have hoped this could be the beginning of a series of such texts, but this seems to have been the only one produced. In the 1960s, the "crazy" often was less the theories than the trajectories, as he responded to the knock of opportunity more than to the potential developments in his own field of expertise. Like an English professor writing a book on a sexologist?

This free-range scholarship also went in even more surprising directions, however. In 1965, Money published an article on "Negro Illegitimacy" (1965c), and in 1969 he gave a talk to a Baltimore library called "Understanding the Ghetto," in which he made some comparisons between African-American life in Baltimore and Aboriginal life in Australia (1969f). Money's constant references to the sexual life of Aboriginal communities in Arnhem Land depended on very slim evidence, and his right to speak on "the ghetto" seems to have been based on the house he owned there. Michael King (1998) wrote of Money's commitment to live in a run-down row house in an African-American neighbourhood near the hospital and the ambivalent respect "Mr. Doc" received from his neighbours, but it is difficult to see this as justification for the lecture.

While the Australian and African-American references were not very important parts of Money's work, they were representative of his easy assumption of authority. This, I presume, is also the explanation of his various letters to American senators. A number were on biomedical research and other such matters relevant to his work, but many were not. Just as one example, in 1973 he wrote a number of letters, on the stationery of the psychohormonal unit, to Senator John Glenn Beall on an issue that seems to have little to do with his expertise – the state of the American presidency: "I am in favor of instituting impeachment against Nixon without delay, in order to forestall his taking over by way of a military coup" (in JMCK). The letter was signed "John Money, PhD Professor of Medical Psychology and Associate Professor of Pediatrics." The idea seems at best strange, but Money's assumption that it was acceptable to write the letter in his professional capacity seems to be an egregious misuse of his position.

I found some insights into what is wrong with Money's assumptions of his sweeping authority by looking at the work of Stephen Muecke, an Australian professor of cultural studies. Muecke is a respected outsider in one Aboriginal community. He has written often of the difficulty of learning and sharing

Aboriginal culture, as he does in "A Chance to Hear a Nyinga Song" (2008). Muecke describes the empathy required even to enter the community: "The experience of being a visitor, or 'training the imagination to go visiting,' is one of imagining oneself in the place of the Other while remaining oneself" (85). If one achieves this, then perhaps one can learn something that can be taken away:

> But of course the ritual event and its repetition would not end the story, which has another paradigm, that of ongoing decolonization which requires more than the ritual installation of feelings (appreciation), it requires all the necessary detail for understanding history and the emerging place of different histories as a precondition for a society seeking justice. (93)

In this approach, Muecke shows the recognition of impossibility that must accompany the transmission of some aspect of this other culture, long before having the temerity to analyze that aspect.

Perhaps Money's failure in each of his fields was such a failure of recognition of the full self of the other. His sympathy was manifest throughout his work. While he seldom seems to have had great respect for his colleagues, he always had it for his patients. However, his idea of justice was never of the need for empathy but rather of the need for a rational assessment of the case, an assessment that often led him to decisions that others – and "the Other," in this instance the patient – might not have thought to be the best one. Money's various comments on Australia, Indonesia, and India invariably assert a view of the culture that exists only to reinforce a view of American sexuality that Money already had. He was not "in the place of the Other"; rather, he saw "the Other" as an object that might aid his pursuit of sexual science and that he might aid.

Recent work in sexology offers none of the freewheeling thought characteristic of Money and of a few others of his generation, such as Albert Ellis. They saw sexology as not a broad field in which a professional might have a small slice of expertise, in Money's case as a pediatric psychoendocrinologist. Instead, defining oneself as a sexologist offered the opportunity to think about sex and gender in the widest possible way. The field itself seemed entrepreneurial: the boundaries of its tangents were as limitless as the imagination of the sexologist, and it was up to the energy of the sexologist to push those boundaries as far as possible. "Empirical checking" might be important, but it was quite possible for it to be reduced to lip service.

Money's work showed a consistent belief in the possibilities of science, even when he was, in a sense, opposing science. This is often evident in his language, especially in his early work. Thus, for the book *The Sex Chromatin* in 1966, Money wrote one chapter titled "The Sex Chromatin and Psychosexual Differentiation." He concluded, "A discrepancy of psychosexual differentiation may occur not only in hermaphrodites but also in morphological normal persons, producing a disparity between the psychosexual identity and the sex chromatin pattern; in extreme degree this becomes a complete contradiction" (1966b, 441). To have said simply that transsexuals have normal chromosomes would no doubt have seemed just too simple and too unscientific. And yet, at the same time as he believed wholeheartedly in science, he saw many ways in which science failed him. In the transcript of the interview for *Omni* magazine in 1986, Money was still very angry that the Meyer study had caused the end of gender reassignment surgery at Johns Hopkins, and he dismissed the possibility that the study represented science: "I told you already that science has got nothing to do with truth; it's got to do with fashion, in many instances" (1986a, 34). Money admitted, however, that the truth in medicine is not always an easy thing to ascertain: "So, since medicine is not an exact science, and is based to a very high degree on presuppositions and making the best decision you can in the face of being bamboozled, then you constantly meet these diametrically opposed points of view, from different specialists in different sub-species of medicine" (35). Medicine becomes like sexology, a space of great possibility, but with those possibilities came incessant power struggles.

Money admitted that, in the hierarchy of science, medicine has its limitations:

> None of the animal experimental people in the United States would ever dare mention anything about the applicability of their experiments to human beings in case they lost prestige in the animal experimental world. Because there's a great divide between the real scientists and the phony ones, you see ... the phony ones work with human beings and do clinical work, so they don't control their variables, you see. Of course, the rat people don't either, but they can't see that. (1986a, 146)

Money used animal studies when they supported what he wished to claim, but he also dismissed them when they seemed to prove something that opposed him. I have seen no suggestion that he ever did any research work with animals. The mechanical process of most of the studies done by "rat people" would require a methodical approach quite out of keeping with Money's very subjective relationship with his human subjects. Much of his best work was produced not by being methodical but by being rather freewheeling and receptive, as he asked his patients questions that seemed surprisingly direct but drew out similarly direct answers. So much of Money's work allowed the patient an independence of thought equal to his own. His papers include a number of transcripts in which the responses range far from anything that could be addressed in a coherent study of a specific problem. He could not control his variables and perhaps did not really want to. He left the phony control to the rat people.

There have been many "crises of science," too many to be enumerated. One was marked in 1985 by the publication of *Leviathan and the Air-Pump* by Steven Shapin and Simon Schaffer, which questioned the validity of the claims to objectivity in the experimental method. As the title might suggest, Shapin and Schaffer were more interested in the hard sciences, physics and chemistry, but they were writing about the attitudes Money attributed to the rat people. At the same time as he was so dismissive, he was compelled by this model throughout his life. Both in his more narrow field of pediatric psychoendocrinology and in the larger one of sexology, he was compelled to create a "science" as in the definition provided by the *OED*: "A branch of study which is concerned either with a connected body of demonstrated truths or with observed facts systematically classified and more or less colligated by being brought under general laws, and which includes trustworthy methods for the discovery of new truth within its own domain." I suggested above that Money seemed to believe his "domain" to be limitless. Perhaps it would be better to say that the limits remained to be defined and that Money felt free to define them. His assessment of what were "general laws," and his belief in his own "trustworthy methods," convinced him of "new truth" that he did not sufficiently question. If he was convinced it was science, it was science.

Look at all the problems of the Reimer case. Reimer's own problems were of course personal, but the problems of methodology that Colapinto lays at Money's door are connected to this view of science. There are similar problems in the work of Colapinto's hero, Milton Diamond, with his various assumptions of what biology dictates; however, these were not the problems that harmed Reimer or the problems that defeated Money. As one example, Diamond has often claimed that the Reimer case was "an experiment." This no doubt unfair assumption, as Colapinto points out, connects neatly with Money's constant assertion – an assertion that can easily be read as a lament

– that ethics preclude experimenting with humans, that the only viable way of exploring various issues is through "experiments of nature." The central scientific question was whether a male with normal chromosomes and born with normal genitalia could be brought up as a girl.

Of course, the primary question raised by most observers is whether such a male *should* be brought up as a girl. Money's argument was a clinical one, that the child would lead a better life if brought up as a girl. As in most clinical claims, this has limited value as a general law. In many cases, the broad expanse of human variables allows almost anything to happen. But the scientific question, that "could," might be more limited. Here Money was again following the scientific method, as something proven in one case might be tested in another. The intersex experience suggested this gender reassignment could work. Then there is simply the question of whether the absence of a penis precluded someone being accepted as a male. While this might seem obviously a social question, the whole discussion of the measurement of the micropenis created the impression that the penis is a statistical question, with all the objectivity of quantification.

The micropenis might seem at most a minor part of this discussion, but the judgment by measurement is an important contribution to the possibility of seeing sexology as a science and of seeing the treatment of Reimer as scientific. As Ian Hacking (1990, 7) states in *The Taming of Chance,* the "statistical example" has a circular relationship with "the style of reasoning": "We cannot justify the style as the best way to discover the truth of the proposition, because the sense of the proposition itself depends on the style of reasoning by which the truth is settled." Numerical assessment seems to move the theory to a valuable fact and to part of that "new truth within its own domain." Money's letters from Dunedin suggested that he was very concerned from his early years that psychology be treated as a science. Throughout his career, he used medical science to justify his own theories and clinical practice. He came to believe that, in spite of the many untruths purveyed by science, if pursued properly, science is true. This seems to have been at least part of the reason for the errors in his various pursuits. His obsession with language was something akin to that belief in quantification: he knew that, if he could just define everything precisely, the answers would result. And real answers, scientific answers. These words could transmit the hard answers of scientific sexology, not the soft, philosophical answers of sexosophy.

While many of those who contributed to obituaries of Money, such as Richard Green and Anke Ehrhardt, asserted the importance of his wide-ranging theoretical brilliance, this is not the direction of sexology today.

Instead, the papers given annually to the Society for the Scientific Study of Sexuality tend to be quantitative, objective, based on models that claim the scientific method that cannot be doubted. They tend to ask small questions and receive small, quite dependable, answers. They are not adventurers like Money, exploring the global theories of terra incognita. Of course, part of the reason for this is simply financial. In the *Omni* interview, Money said,

> You know, you can do things when you've got money, and if there were an endowed chair with a big endowment for research, you'd be amazed at how quickly the medical establishment in this medical school would think it was just the greatest thing ... But if the money were there, sex would suddenly become perfectly okay at Johns Hopkins; it really would. (1986a, 153)

If there is a general warning in Money's career, it is no doubt that faith in science, like faith in anything, can lead to blindness. Money was convinced by science and convinced himself of the validity of his own science. The "empirical checking" often was not done; in any case, Money should have had more fear of empiricism. As Hacking (1990) points out, theories seldom arise from experimentation; rather, experimentation is shaped to prove theories. The empirical data that scientists claim as proof are rather a quantitative manipulation of existing theories. But the warning from Money's personal life is perhaps quantitative in another direction.

With the exception of one anti-Semitic letter written to him (a bit of a joke given Money's intensely Christian forebears), Money's own name seems almost never to have been a metaphoric turn for his critics. And yet money was his constant concern, as in the *Omni* comment. Growing up in poverty led him to the absurd frugality so often noted by his colleagues. The house that kept him in "the ghetto" was of course very cheap to buy and was worth almost nothing when he died. His habitual assertion that he was the only researcher to receive constant funding from the National Institute of Child Health and Human Development represented scholarly credibility, but it also represented financial stability. The same could be said of his entrepreneurial excursions into dyslexia. Like many who grew up poor, he never believed that financial stability was possible. Thus, approval by the financial system of "big science" on which he touched in *Omni* was both a reflection of the success of his theories and the one possible source of his personal maintenance. Money's recognition that his theories were avant-garde made him feel financial support was always endangered, and yet the energy of his career seems to have been based on his own sense of himself as a radical theorist of sex and gender. He always thought both ideas and income were constantly under attack, by outside forces but also by each other. The contradictions of this position no doubt contributed to his tendency to feel threatened and to react belligerently to opposition.

The second chapter of this book shows some of the elements of Money's upbringing that shaped his character. While this is not a biography, his youth offers many elements worthy of note given his accomplishments as a sexologist. I suggest above that his poverty was no doubt a constant part of his approach to research. The other major difficulty of his childhood, the violence of his father and then his early death, provides many innuendoes about the stimulus for Money's work on masculinities and on childhood; while these are often interesting, they remain innuendoes rather than causes. I would place more emphasis on his experience as a New Zealander. New Zealanders are always aware that they are from a small place, far away from any location that is usually regarded as internationally important. One example is a T-shirt that reads "London Paris Rome Ponsonby," the last being a suburb of Auckland, New Zealand's largest city. The irony is clear, but then I add to it with that explanatory note about Auckland. Presumably, no one would feel the need to explain "London, England's largest city."

The irrelevance of New Zealand has meant that only New Zealanders have ever expressed much interest in Money's origins. *As Nature Made Him* begins with the description of him as "British-sounding," which might have been the impression he made on the Reimers but certainly misrepresents the incessantly colonial position of a New Zealander. A more acute ear might have noted the specifics of a New Zealand pronunciation: perhaps Money chose his field because he came from a country in which "sex" is the number after five. Money's push for the avant-garde in sexology no doubt partly responds to the religious repression of his youth, but it also probably responds to what he always felt were the limitations of his New Zealand education. As a person from nowhere, his accomplishments somewhere, in the United States, meant a great deal. When I mention him in a lecture, I usually refer to him as an "American sexologist," which is strictly incorrect but probably would please him.

Money's qualifications as an American sexologist were established by his dissertation on intersex. Both the substance of the dissertation and the resulting appointment at Johns Hopkins made him an early innovator. His depiction of the intersex individuals in his study as what might be called "normal" became central to his work throughout his career, as he looked at a variety of gender and sexual diversities and suggested ways in which they could be accepted and accommodated and understood. Money saw the determining of intersex individuals as psychologically normal as suggesting the need to make intersex infants as specifically gendered as possible, so they could more easily live that normal life. This included genital surgery to establish to them and to those caring for them that they were normal. Money's ability to turn this into "the Money protocol" provided him with the professional power he needed for his many other endeavours, but it also made him a central target for those who opposed his procedure. If the individual is normal, why is there any need for surgery? The potential contradictions in Money's argument led some to believe that his commitment to medical procedures and to scientific theories trumped his concern for the humans involved. This meant that in his fifty years in the field his reputation moved from being the agent of progress to someone often attacked as mired in past procedures that lacked respect for the intersex individual. The result is not only the rejection of Money by the intersex community but also a much more wide dismissal by most aspects of sexuality and gender studies.

While there can be no question that it was Money's work on intersex that established his credentials, and that it is certainly possible that his theories on intersex might have the most long-lasting effect, he devoted most of his publication energy to the widest possible view of sex and gender. He saw no aspect of human life not shaped by sex and gender, and he seemed to have felt little hesitation to comment on all of those aspects. His position at Johns Hopkins gave him the opportunity to examine the whole of the early lives of intersex individuals. *Intersex* in this case meant individuals with sufficient endocrinological problems to be referred to Johns Hopkins. The sample was not large, but, because he saw them as normal, he recognized the unusual opportunity for a sexologist to examine the whole of development of a normal person. When they exhibited the usual percentage of problems and dysfunctions, he used their cases for theories in a variety of directions. One central element was what are now called paraphilias. Using these as a template, he came to understand how the whole of a psychosocial life could be understood in terms of a psychosexual development or, in his terms, a "lovemap." While the term has not been used widely, the idea of understanding any individual through one developmental template, one that understands everyone from the most asexual to the criminal sexual deviant as a product of a certain process, continues to be attractive to many theorists and practitioners. I do not mean to suggest any direct influence from Money, but various concepts that today seem generally accepted are based on a continuum that incorporates the sexuality of all humanity. A phrase such as "sexual diversities," often used to define academic programs, implies an overall understanding that can incorporate homosexuality, transgender, intersex, and any other sexual and gender possibilities. Even the contemporary usage of *queer* fits this template, as it so often includes people who in an earlier era would have been "just" heterosexual but now see themselves as outside the traditional division between "normal" and other. Yet there also continue to be examples of specific models that reflect Money's paradigms. In conversation, one psychologist who works with incarcerated pedophiles, and who actually had little knowledge of Money's work, described his understanding of his patients in a way that almost perfectly repeated Money's lovemap theory. What remains the standard text for the treatment of pedophiles, Berlin and Krout's article "Pedophilia" (1986), was shaped by Money's concepts.

Money's research base was with intersex patients, but his fame was often associated with transsexuals, particularly through the clinic he established at Johns Hopkins. This is never more obvious than in Janice Raymond's attack on transsexuals ([1979] 1994), where she singles out Money as the prime example of someone who feeds the delusions of men who want to be women. While her representation of Money is inaccurate, she is correct in seeing him as the person in the late 1970s who was the most public medical supporter of trans women. While he had various theories of the possible causes of transsexuality, this is a particular field in which his general dismissal of assumptions of causality had its greatest power. He believed that transsexuals could not be convinced that their gender contradiction was a delusion and, thus, whatever the cause and whatever society's view of their situation, the only responsible medical reaction was to accept what he termed their idée fixe and accept their self-diagnosis and offer them the surgery they requested.

Money's work on homosexuality is intriguing for many reasons. One is simply that it must have been in some sense personal. His own sex with men was common knowledge among his colleagues, but I have seen no suggestion that he was personally involved in any other aspect of the sexual or gender diversities. Second, while his experience of homosexuality, introspectively, practically, and socially, was extensive, his research in the field was only inadvertent. This did not prevent him from writing a book on the subject and from acting as an expert on homosexuality in a number of court cases. Once again, his theories of causality were at best vague, and he seems to have found them of limited importance. Instead, his assessment was not unlike Freud's response to homosexuality: homosexuality is a fact, and nothing any psychologist, surgeon, or policeman does is going to change that. The psychological problems experienced by homosexuals are the result of homophobia, in law and in society in general, not the result of sexual orientation. He spoke out again and again on the need to remove restrictions on homosexuals in the military, in education, and in all aspects of life. He was particularly concerned that the education of children should include knowledge of homosexuality. His stated position was that such education was likely to create more healthy heterosexuals, but given the historical context it might be reasonable to interpret this as meaning more healthy individuals of all sexualities.

Money's obsession with words raises an interesting question about causality: why? I have no convincing answer. However, it seems to have been a major part of his intellectual endeavours throughout, visible even in his early letters in New Zealand. Still, at least a partial explanation might be the importance and power of the way he framed *gender* in his dissertation. In almost all of his early writing, this concept and his own understanding of its meanings are central. His belief in his achievement at creating this word, regardless of possible questions as to how much he actually created it, led to his belief in the power of his own sexological research and the power of words, particularly in taxonomy. Again and again, he attempted to establish links and descriptions, through neologisms, lists, and different forms of "maps." The drive to prove through science became less a yearning for experimentation, as one might expect, than a yearning for definition, to harness the power of language to establish an understandable object, of gender, of sexuality, or of sex. The more I explored Money's texts, the more I recognized the importance of his words, to him and to the culture of the late twentieth century. The infamy with which his career ended cannot erase the great number of people he convinced with those words.

One of whom was Hugh Hefner. It was Money's words that led Hefner to present him in *Playboy*. As a result of his ability to present ideas in a form that appealed to the public and to the public media, Money was ubiquitous from the mid-1970s to the '90s. The effect was to make his ideas much more generally known than those of other sexologists and to make him the resource person for anyone researching any aspect of sexuality or gender, from a convict serving time to a syndicated advice columnist. Money saw this as part of his role as an educator, but it also assisted his constant proselytizing for sexual liberation. His prominence and the desire to be avant-garde became circular. It was not that he did not believe in radical views of sex and gender but rather that he recognized that it was often his radicalism that attracted the press, and he found the press to be very useful to purvey his ideas, which were often radical. The result was to make him famous but also vilified by conservatives and subject to the type of caricature of him found in Jeffrey Eugenides's novel *Middlesex* (2003).

While Money's views were becoming more and more subject to attack by conservatives, and while conservatism gained more and more of a hold on the American imagination, a specific case in which Money was involved came to the fore. Regardless of the details, the barebones of the plot of the Reimer case doomed Money: he had advised that a male infant could be brought up as a girl, and this had failed. It became an opportunity for biological determinists to attack the whole idea of social construction. Money could claim, as he often did, that he was not a "social constructionist" but rather an "interactionist," someone who believed that the self is a result of many biological and environmental forces that combine in different ways at different times of a life, from conception – and even before – to death. Still, this didn't matter. The world quickly saw him as a mad scientist who believed gender to be completely malleable and tried to experiment on a poor defenceless child. Given that a significant part of that world was already uncomfortable with Money's radical openness to sexuality, his support for transsexuals, his defence of pedophiles, and his testimonies for homosexuals, he was an easy target, as *As Nature Made Him* shows.

Money was the founder or co-founder of many associations and journals. He was a constant initiator of projects. Still, many or perhaps most of his publications were either in minor journals or in journals just starting out that had yet to become significant. There could have been many reasons for this, including his desire to help fledgling or weak operations. Still, it is tempting to wonder whether he made this choice to avoid the kind of editorial assessment that made him so angry at Richard Green. In any case, his support and commitment to these various operations have seldom led to an ongoing respect, except from a small coterie that includes Green, Bullough, Lehne, and a few other colleagues, mostly his former students. For the most part, they are retired or semi-retired and seem to have had little influence on Money's reputation.

One possible exception is Eli Coleman, director of the Program in Human Sexuality at the University of Minnesota and a significant figure in sexuality studies for many years. He edited *John Money: A Tribute* (1991) and has continued to do what he can to maintain the knowledge of Money's importance. Through him, Money's family established the John W. Money Endowed Fund at the Program in Human Sexuality, which provides funding for the

John Money Lecture in Pediatric Sexology lecture series hosted at the Program in Human Sexuality, the John Money Plenary at the Biennial World Conference of Sexual Health, and the John Money Award for sexology research, sponsored by the Eastern Region of the Society for the Scientific Study of Sexuality.

The 2011 annual meeting of the Society for the Scientific Study of Sexuality, of which Money was a founder and later president, is possibly more representative of the society's general attitude than Coleman and the awards. One of the plenary speakers at its annual meeting in Texas was Milton Diamond. His topic was "What Can We Learn from the John/Joan Case? Biased Interaction Theory." In an email correspondence with me, Dr. Diamond said, "Much of Dr. Money's ideas were responsible for engendering wrong thinking and bad medicine associated with the management of trans and intersex conditions. I am trying to rectify those wrongs." I hope this book notes much of that "wrong thinking," but it also notes much that is insightful and even, dare I say it, "right thinking."

When Camille Paglia claimed in 1993 that Money was "the leading sexologist in the world today," she was asserting a category that would seem no longer to exist. It would be difficult to think of someone who fit that phrase today. One might posit Anne Fausto-Sterling, a biologist, or Cordelia Fine, a psychologist, both of whose work I use in the present study, but neither fits the mould of Kinsey or Money. The most significant participants in the Society for the Scientific Study of Sexuality do not have the public prominence that sexologists once enjoyed. Those most praised within the organization tend to be those with precision and objectivity that are quite unlike the opinionated Dr. Money. On the other hand, those with the largest public profile on sexual issues tend to be those who are not involved in medical research or treatment; rather, they are journalists who write trade books, such as Naomi Wolf, author of *Vagina: A New Biography* (2012), or Hanna Rosin, author of *The End of Men* (2012). Some of those who are very involved in Money's areas of study are often quoted on sexuality and gender issues, such as Judith Halberstam, author of *The Queer Art of Failure* (2011), and Dean Spade, author of *Normal Life: Administrative Violence, Critical Trans Politics and the Limits of Law* (2012), yet it is difficult to see why they would be called "sexologists." Above I suggested conservatism as part of the reason Money's ideas went out of fashion, but there might also be a general demise of the sexologist as a person with specific medical expertise who could make pronouncements on a variety of sexual and gender issues. When authors such as Halberstam and Spade speak, the expertise is a combination of the personal and the academic, Halberstam as a professor of literature and Spade as a professor of law. There is no claim to medical expertise, nor any great interest in that expertise. The question is how identity is at play. Money was concerned with identity, and his advocacy was often in aid of persons with specific identities receiving the medical care that he perceived them to need. But his claimed expertise was as a medical psychologist and sexologist, and he never spoke in an identity other than that.

The atmosphere seems in many ways a somewhat strange combination of what Money worked towards and its opposite. In terms of the former, there is a general acceptance within the sexual diversities of all being normal. In other words, those with various gender or sexual identities other than what society as a whole expects can be accepted as functional, persons who have no need for medical or other intervention except that which they choose to have. This is exactly how Money hoped to treat all patients, including those who are transgendered. The one area of his approach to transsexuals that is now rejected by much of the transgendered community is the real-life test, in which a person is required to live in the opposite gender for one or two years in order to prove the need for reassignment surgery. Money saw this as fulfilling the need to prove that this could be a normal condition. There was a similar impetus in his commitment to surgical interventions for intersex infants, to give them the greatest opportunity to perceive themselves to be "normal."

In the early twenty-first century, those involved in various aspects of the sexual diversities see no need for such intervention for an intersex person, and most would see the real-life test for a transgender individual to be something that should be up to that individual. In other words, Money's obvious error, in terms of contemporary views, was to believe too strongly in the medical gatekeeper. This was a major mistake in the Reimer case, which could have been handled with much more openness to gender possibilities. Money certainly believed that the gender of an infant could be changed, but he also believed that it need not be and that in certain cases a gender reassignment could be reversed, as Reimer himself wished. That this reversal was not aided by Money continues to be the error that has most savaged his reputation. The day before I wrote this paragraph I was talking to a trans woman who was asking about my book. She immediately became angry at the thought that I might have something positive to say about a medical person who believed in surgery to enforce a gender assignment not wanted by the individual. This is how Money is known.

There is also an opposite problem, primarily about how Money is perceived outside the community of sexual diversities. This is what I would still call the biological determinist model. It is the view created by John Colapinto and *As*

Nature Made Him. At least on the surface, this position seems quite opposite to the sexual diversities world: you are not a person who can choose how to express your gender and sexuality. You are rather someone whose biology ordains what you should be. And yet some of the great believers in such biological determinism are also great supporters of diversity, such as Milton Diamond and Simon LeVay. They accept diversity not because of the freedom of the individual but because of the inevitability of a form of biological determinism that will someday be discovered, which will explain why someone is transsexual or homosexual. Whether the argument is against Money's support of gender and sexual freedom, or against his belief in the necessities of medical intervention to accomplish such freedom, or against his interactionist theories of development, the argument is against Money.

In 2002, he published a brief article on meeting Kinsey in 1954. Judging from the manuscript in his collection, the original seems to have been written in the 1950s. If so, Money did a good job of forecasting his own future reputation through his portrait of Kinsey: "He was the prophet of righteousness being persecuted without cause, quite unable to glimpse himself as others see him, namely as a man who can excel at antagonizing others with his bumptious arrogance, and rancorous disparagement" (2002b, 321). As Money extended his analysis of Kinsey, it became more and more revealing of Money himself:

> He is not one of those research men who can go unobtrusively about his work, intrigued by his own findings but indifferent to spreading the news of them abroad. If he were, he would have recognized, right from the start, that the civilization which invented science and the idea of research happens also to be a civilization which is prudish and hypocritical about human sexual behavior, and that as a member of this civilization doing research he has to put up with the prudishness, bucking it as little as possible. Whether he knows it or not, and whether he likes it or not, he is bucking the prudishness not as little, but as much as possible. As the prophets denounced the errors and wickedness of their people, and as a Copernicus or Galileo proclaimed the errors of religious theories of the universe, Kinsey is intent on revealing the falseness and hypocrisy of our sexual beliefs and customs. Not only will he study the sexual behavior of men and women, but he will change it too. Not only is he scientist, but prophet and preacher also, and the prophet is often without honor among

> his own people, for human beings resent and resist change as much as they court and espouse it. (322)

In the end, Money's generalization in the 1950s seems to have recognized a category and its characteristics: "What I did leave the symposium with was a perplexity – a perplexity whether the men who have great impact on society and its ways must always be peremptory, intolerant prophets who engender self-effacing loyalty in their acolytes, antagonism and hostility in their persecutors, and impartial spectatorship in almost no one" (322).

When I was writing *Queersexlife,* Money was an inspiration and a frustration. He was an inspiration because of the independence and brilliance of his ideas and his willingness again and again to be on the side of sexual liberation. He was a frustration because his errors so often seemed to be a product of arrogance and a rather willful blindness. No other sexologist offered so many brilliant and insightful theories and wrote so much that is so lacking in awareness. In the Reimer case, that lack led to destruction, of David Reimer but also of John Money. The summary of Money's career is certainly not "as nature made him," but it might be "as Money made himself." Thus, the tragedy is that the self that created was perhaps inevitably the self that enabled the destruction.

Works Cited

Archival Sources

Canadian Lesbian and Gay Archives, Toronto.
John Money Collection, Hocken Library, University of Otago [JMCH].
John Money Collection, Kinsey Institute, University of Indiana [JMCK].
Michael King Collection, Turnbull Library, Wellington, New Zealand [MKCT].

Works by John Money

(in chronological order)

Abbreviations

VP	Articles reprinted in *Venuses Penuses* (page numbers from that publication are included)
JMCK	Unpublished material from the John Money Collection, Kinsey Institute, University of Indiana
JMCH	Unpublished material from the John Money Collection, Hocken Library, University of Otago
MKCT	Material from the Michael King Collection, Turnbull Library, Wellington, New Zealand

n.d. "Psychiatry and Religion." Unpublished paper, University of Otago. JMCH.
1944a. "Creative Endeavour in Musical Composition." Unpublished MA thesis, University of Otago.
1944b. "The Social Scientist on Freedom." *Spike: The Victoria University College Review* 43, 7: 25-26.

1945. "Delusion, Belief and Fact." Unpublished paper.
1945-46. Various letters to his family from University of Otago. JMCH.
1947. "Basic Concepts in the Study of Personality." Paper presented to British Association for the Advancement of Science Congress, Wellington, May 22. JMCH.
1948. "Delusion, Belief and Fact." *Psychiatry* 11: 33-38. *VP*, 21-29.
1949a. "A Specific Problem in Methodology: An Example." January. JMCK.
1949b. "Unanimity in the Social Sciences with Reference to Epistemology, Ontology and Scientific Method." *Psychiatry* 12: 211-21. *VP*, 30-44.
1952. "Hermaphroditism: An Inquiry into the Nature of a Human Paradox." PhD diss., Harvard University.
1954. "The Diagnosis and Treatment of Endocrine Disorders in Childhood and Adolescence." [Written in 1954 for Lawson Wilkin's book of the same title. No further information given.] JMCK.
1955a. "A Children's Medical Book of Sex" (also includes letter of rejection). JMCK.
1955b. [and Joan G. Hampson, and John L. Hampson]. "An Examination of Some Basic Sexual Concepts: The Evidence of Human Hermaphroditism." *Bulletin of the Johns Hopkins Hospital* 97, 4: 301-19. *VP*, 152-71.
1955c. [and Joan G. Hampson, and John L. Hampson]. "Hermaphroditism: Recommendations Concerning Assignment of Sex, Change of Sex, and Psychologic Management." *Bulletin of the Johns Hopkins Hospital* 97, 4: 284-300. *VP*, 133-50.
1955d. "Linguistic Resources and Psychodynamic Theory." *British Journal of Medical Psychology* 28, 4: 264-66. *VP*, 69-72.
1956a. "Mind-Body Dualism and the Unity of Bodymind." *Behavioral Science* 1, 3: 212-17.
1956b. Review of *Sexual Behavior in American Society: An Appraisal of the First Two Kinsey Reports*, edited by Jerome Himelhoch and Sylvia Fleis Fava. *Contemporary Psychology* 1, 8: 232-34.
1957. [and Joan G. Hampson, and John L. Hampson]. "Imprinting and the Establishment of Gender Role." *AMA Archives of Neurology and Psychiatry* 77: 333-36. *VP*, 186-90.
1960a. "Components of Eroticism in Man: Cognitional Rehearsals." In *Recent Advances in Biological Psychiatry*, edited by J. Wortis, 210-25. New York: Grune and Stratton.
1960b. [and Richard Green]. "Incongruous Gender Role: Nongenital Manifestations in Prepubertal Boys." *Journal of Nervous and Mental Disease* 130, 8: 160-68.
1960c. "Phantom Orgasm in the Dreams of Paraplegic Men and Women." *Archives of General Psychiatry* 3: 373-82.
1961a. "Components of Eroticism in Man: I. The Hormones in Relation to Sexual Morphology and Sexual Desire." *Journal of Nervous and Mental Disease* 132, 3: 239-48.
1961b. "Components of Eroticism in Man: II. The Orgasm and Genital Somesthesia." *Journal of Nervous and Mental Disease* 132, 4: 289-97.
1961c. [and Richard Green]. "'Tomboys' and 'Sissies.'" *Sexology* 28, 5: 2-5.
1961d. "Too Early Puberty." *Sexology* 28, 3-4: 154-57, 250-53.
1962a. "Obstacles to Sex Research." *Sexology* 28, 8: 548-53.
1962b. *Reading Disability: Progress and Research Needs in Dyslexia*. Baltimore: Johns Hopkins University Press.

1963a. [and Steven R. Hirsch]. "Chromosome Anomalies, Mental Deficiency, and Schizophrenia: A Study of Triple X and Triple X/Y Chromosomes in 5 Patients and Their Families." *Archives of General Psychiatry* 8: 242-51.

1963b. "Developmental Differentiation of Femininity and Masculinity Compared." In *Man and Civilization: The Potential of Woman,* edited by S.M. Farber and R.H.L. Wilson, 51-63. New York: McGraw-Hill.

1963c. "Factors in the Genesis of Homosexuality." In *Determinants of Human Sexual Behavior,* edited by George Winokur, 19-43. Springfield, IL: Charles C. Thomas.

1963d. Letter to Isadore Rubin, July 9. JMCK.

1963e. "Psychosexual Development in Man." In *Encyclopedia of Mental Health,* edited by Albert Deutsch and Helen Fishman, 1678-1709. New York: Franklin Watts.

1964a. Review of *Homosexuality: A Psychoanalytic Study of Male Homosexuals,* by Irving Bieber et al. *Journal of Nervous and Mental Disease* 138: 197-200.

1964b. [and Ernesto Pollitt and Steven Hirsch]. "Priapism, Impotence and Human Figure Drawings." *Journal of Nervous and Mental Disease* 139, 2: 161-68.

1965a. [and Steven Hirsch]. "After Priapism: Orgasm Retained, Erection Lost." *Journal of Urology* 94: 152-57.

1965b. "Interview with Maxine Davis." September 4. JMCK.

1965c. "Negro Illegitimacy: An Antebellum Legacy in Obstetrical Sociology." *Pacific Medicine and Surgery* 73: 350-52.

1965d. "Psychology of Intersexes." *Urology International* 19: 185-89.

1965e. [and Duane Alexander]. "Reading Ability, Object Constancy, and Turner's Syndrome." *Perceptual and Motor Skills* 20: 981-84.

1965f. "Scientific Man." Unpublished talk given to a parents' group at the Park School, Baltimore, May 13. JMCK.

1965g. "The Sex Instinct and Human Eroticism." *Journal of Sex Research* 1, 1: 3-16.

1965h. *A Standardized Road Map Test of Direction Sense.* Baltimore, MD: Johns Hopkins University Press.

1966a. [and Christine Wang]. "Human Figure Drawing. 1: Sex of First Choice in Gender-Identity Anomalies, Klinefelter's Syndrome and Precocious Puberty." *Journal of Nervous and Mental Disease* 143, 2: 157-62.

1966b. "The Sex Chromatin and Psychosexual Differentiation." In *The Sex Chromatin,* edited by Keith L. Moore, 434-43. Philadelphia: W.B. Saunders.

1966c. "Which Sex Doth Prevail?" *Journal of the American Medical Association* 196, 5: 447.

1967a. "Cytogenic and Other Aspects of Transvestism and Transsexualism." *Journal of Sex Research* 3, 2: 141-43.

1967b. "Progress of Knowledge and Revision of the Theory of Infantile Sexuality." *International Journal of Psychiatry* 4: 50-53. *VP,* 530-34.

1967c. "Sexual Problems of the Chronically Ill." In *Sexual Problems: Diagnosis and Treatment in Medical Practice,* edited by Charles W. Wahl, 266-87. New York: Free Press.

1967d. "25 Years a Man, Now 15 Years a Woman." *Baltimore Sunday Sun,* September 17.

1967e. [and Ralph Epstein]. "Verbal Aptitude and Prepubertal Effeminacy: A Feminine Trait." *New York Academy of Sciences* 29: 448-54.

1968. "Hermaphroditism and Related Incongruities." Written for *Encyclopedia of Homosexual Behavior,* which never went to press. JMCK.

1969a. [and Florence Schwartz]. "Public Opinion and Social Issues in Transsexualism: A Case Study in Medical Sociology." In *Transsexualism and Sex Reassignment,*

edited by Richard Green and John Money, 253-66. Baltimore, MD: Johns Hopkins University Press.
1969b. "Sex Education." Unpublished lecture to University of Maryland Medical School, February 4.
1969c. [and Clay Primrose]. "Sexual Dimorphism and Dissociation in the Psychology of Male Transsexuals." In *Transsexualism and Sex Reassignment,* edited by Richard Green and John Money, 115-31. Baltimore, MD: Johns Hopkins University Press.
1969d. [and John G. Brennan]. "Sexual Dimorphism in the Psychology of Female Transsexuals." In *Transsexualism and Sex Reassignment,* edited by Richard Green and John Money, 137-52. Baltimore, MD: Johns Hopkins University Press.
1969e. [and Reynolds Potter and Clarice S. Stoll]. "Sex Reannouncement in Hereditary Sex Deformity: Psychology and Sociology of Habilitation." *Social Science and Medicine* 3: 207-16.
1969f. "Understanding the Ghetto." Unpublished presentation. March 3. Randallstown Library, JMCK.
1970a. "Critique of Dr. Zuger's Manuscript." *Psychosomatic Medicine* 32, 5: 463-64.
1970b. [and John G. Brennan]. "Heterosexual vs. Homosexual Attitudes: Partners' Perception of the Feminine Image of Male Transsexuals." *Journal of Sex Research* 6, 3: 193-209.
1970c. "New Direction: Psychoanalytic Libido Theory." *Contemporary Psychology* 15, 3: 226-27.
1970d. "Paraphilias: Phyletic Origins." Lecture for Introduction to Psychiatry and Behavioral Sciences, Johns Hopkins University. JMCK.
1970e. "Paraphilias: Phyletic Origins." Lecture for Introduction to Psychiatry and Behavioral Sciences, Johns Hopkins University. JMCK. *VP,* 454-79.
1970f. "Sex Reassignment." *International Journal of Psychiatry* 9: 249-82.
1970g. "Sexual Dimorphism and Homosexual Gender Identity." *Psychological Bulletin* 74, 6: 425-40.
1970h. "Significant Aspects of Erotica." Unpublished typescript marked "1970." JMCH.
1970i. Teenage Interviews. In "Pornography in the Home and in Medical Education" Binder. JMCK.
1971a. "The Aboriginal Australians of Arnhem Land and Psychosexual Differentiation in Childhood." In *Sex in Childhood,* 7-20. Tulsa, OK: Children's Medical Center.
1971b. [and Barry A. Kinsey, James T. Proctor, and René Spitz]. "First Panel Discussion." In *Sex in Childhood,* 65-70. Tulsa, OK: Children's Medical Center.
1971c. "Pornography in the Home." In *Sex in Childhood,* 71-83. Tulsa, OK: Children's Medical Center.
1971d. Review of *The Intersexual Disorders,* by Christopher Dewhurst and Ronald Gordon. *Journal of Nervous and Mental Disease* 152: 216-18.
1971e. "Transsexualism and the Philosophy of Healing." *Journal of the American Society of Psychosomatic Dentistry and Medicine* 18, 1: 25-26.
1972a. "Determinants of Human Sexual Identity and Behavior." In *Progress in Family and Group Therapy,* edited by C.J. Sager and H.A. Kaplan. New York: Brunner/Mazel. *VP,* 201-22.
1972b. "Herodotus of Homosexuality." Review of *Sexuality and Homosexuality: A New View,* by Arno Karlen. *Contemporary Psychology* 17, 6: 355.

1972c. Letter to Michael Simpson, April 3. JMCK.
1972d. [and Anke A. Ehrhardt]. *Man and Woman, Boy and Girl: The Differentiation and Dimorphism of Gender Identity from Conception to Maturity.* Baltimore, MD: Johns Hopkins University Press.
1973a. "Biology = Destiny: A Woman's View." Review of *Males and Females,* by Corinne Hutt. *Contemporary Psychology* 18, 12: 603-4.
1973b. "The Birth-Control Age." *Chronicle of Higher Education.* November 12: 20-21.
1973c. "Sexpert Testimony." In *Getting into Deep Throat,* by Richard Smith, 255-79. Chicago: Playboy Press.
1973d. "Unisex, Ambisex, Bisex? The Liberation of Both Sexes." *Johns Hopkins Magazine* 24, 4-5: 4-5.
1974a. "My Orgasm Belongs to Daddy." Review of *The Female Orgasm,* by Seymour Fisher. *Contemporary Psychology* 19, 5: 399-400.
1974b. "Prenatal Hormones and Postnatal Socialization in Gender Identity Differentiation." *Nebraska Symposium on Motivation* 21: 221-95. *VP,* 108-22.
1975a. "Expert Witness Testimony on Homosexuality, in the Case of *L.M. Smith v. Kissinger, U.S. Department of State Agency for International Development,* Washington, D.C. 3.25.1975." JMCH.
1975b. "Nativism versus Culturism in Gender-Identity Differentiation." In *Sexuality and Psychoanalysis,* edited by E. Adelson, 48-62. New York: Brunner/Mazel.
1975c. [and Patricia Tucker]. *Sexual Signatures: On Being a Man or a Woman.* Boston: Little, Brown and Company.
1975d. "Stability and Change in Gender Stereotypes." *Totus Homo* 6, 1-2-3: 41-49.
1976a. "The Development of Sexology as a Discipline." *Journal of Sex Research* 12, 2: 83-87.
1976b. "Gender Identity and Hermaphroditism." [Letter] *Science,* February 27: 872.
1976c. "Sex and Money." *APA Monitor* 7, 6: 10-11.
1976d. *A Standardized Road Map Test of Direction Sense.* Napa, CA: Academic Therapy Publications.
1976e. "Statement on Antidiscrimination Regarding Sexual Orientation." *Journal of Homosexuality* 2, 2: 159-60.
1976f. [and Michael De Priest]. "Three Cases of Genital Self-Surgery and Their Relationship to Transsexualism." *Journal of Sex Research* 12, 4: 283-94.
1977a. [and Russell Jobaris and Gregg Furth]. "Apotemnophilia: Two Cases of Self-Demand Amputation as a Paraphilia." *Journal of Sex Research* 13, 2: 115-25.
1977b. "Bisexual, Homosexual, and Heterosexual: Society, Law, and Medicine." *Journal of Homosexuality* 2, 3: 229-33.
1977c. "Cytogenetics, Behavior and Informed Consent." *Modern Medicine of Canada* 32, 2: 127-30.
1977d. "Destereotyping Sex Roles." *Society* 14, 5: 25-28.
1977e. "Doctor Money Speaks, the Whole World Listens." *News-Letter* [Johns Hopkins University], April 22: 4; and April 29: 1-2.
1977f. [and Daniel P. van Kammen]. "Erotic Imagery and Self-Castration in Transvestism/Transsexualism: A Case Report." *Journal of Homosexuality* 2, 4: 359-66.
1977g. "How Is Gender Identity Formed: An Interview with Dr. John Money." [Interview by Nora Harlow]. *Medical Tribune Sexual Medicine Today,* February 23: 33-37.

1977h. "Introduction Written for 'By Body Betrayed: A Photographic Exploration of Transsexualism' by Jay Hirsch." "Written July 1977." JMCK.
1977i. [and Tom Mazur]. "Microphallus: The Successful Use of a Prosthetic Phallus in a 9-Year-Old Boy." *Journal of Sex and Marital Therapy* 3, 3: 187-96.
1977j. "Issues and Attitudes in Research and Treatment of Variant Forms of Human Sexual Behavior." In *Ethical Issues in Sex Therapy and Research,* edited by W.H. Masters, V.E. Johnson, and R.C. Kolodny. Boston: Little, Brown. *VP,* 535-45.
1977k. "Sex in America Today." *Gallery* 5, 5: 35-37, 56-58, 68.
1977l. "Thalidomide Teens Question." *Sexualmedizin* 6: 962.
1977m. "Tricentennial Time Capsule: The Sexual Counterreformation Documented." Review of *Citizens for Decency: Antipornography Crusades as Status Defense,* by Louis A. Zurcher Jr. and R. George Kirkpatrick. *Contemporary Psychology* 22, 3: 175-76.
1978a. [and Robert Sollod]. "Body Image, Plastic Surgery (Prosthetic Testes) and Kallmann's Syndrome." *British Journal of Medical Psychology* 51: 91-94.
1978b. "Dirt and Dirty in Sexual Talk and Behavior." *MIMS Magazine,* September 15.
1978c. "Gender Identity: A Lesson from Hermaphroditism." Unpublished paper. [Note says written for *Le Fait Feminine* (E. Sullerot, ed.)]. JMCK.
1978d. "Imagery in Sexual Hang-Ups." *The Humanist,* March-April: 13-15.
1978e. "Love." Unpublished manuscript. [Typed note says "written for *Encyclopedia of Psychology*" (eds. W. Arnold, H.J. Eysenck and R. Meili)]. JMCK.
1978f. [and Christopher Jarosz]. "Pedophiliac and Epheboliac Relationships: A Pilot Study." Unpublished, 4 December. JMCH.
1978g. "The Sexual Revolution: A Manifesto." *Forum.*
1978h. "Transsexualism." In *Encyclopedia Americana.* Vol. 27. New York: Grolier's.
1979a. "Freak or Katharma." Review of *Freaks: Myths and Images of the Secret Self,* by Leslie Fiedler. *Contemporary Psychology* 24, 2: 108-9.
1979b. [and Anthony J. Russo]. "Homosexual Outcome of Discordant Gender Identity/Role in Childhood: Longitudinal Follow-Up." *Journal of Pediatric Psychology* 4, 1: 29-41. *VP,* 258-67.
1979c. "Ideas and Ethics of Psychosexual Determinism." *British Journal of Sexual Medicine* 6, 48: 27-32.
1979d. "In the Provincial Courts (Criminal Division) Judicial District of York," *R. v. Pink Triangle Press.* Toronto, January 4, 5, 8. Canadian Lesbian and Gay Archives, Toronto.
1979e. "Sexual Dictatorship, Dissidence and Democracy." *International Journal of Medicine and Law* 1, 1: 11-20.
1980a. "The Birds, the Bees, and John Money." [Interview by Lawrence Mass]. *Christopher Street* 4, 12: 24-30.
1980b. [and Robert Seidenstadt]. "Human Autoerotic Practices." *Journal of Nervous and Mental Disease* 168, 3: 289-97.
1980c. *Love and Love Sickness: The Science of Sex, Gender Difference and Pair-Bonding.* Baltimore, MD: Johns Hopkins University Press.
1980d. [and Peter A. Lee, Thomas Mazur, Robert Danish, James Amrhein, Robert M. Blizzard, and Claude J. Migeon]. "Micropenis. 1. Criteria, Etiologies and Classification." *Johns Hopkins Medical Journal* 146: 156-63.
1980e. "The Need for Sex Rehearsal Play." *Sexology* 46, 9: 21-24.
1980f. [and William R. Cameron]. "Sexism in Psychiatric Nosology and Diagnosis." Review of *Gender and Disordered Behavior: Sex Differences in Psychopathology,*

edited by Edith S. Gomberg and Violet Franks. *Contemporary Psychology* 25, 2: 136-37.
1980g. "*Sexology* Comment on Incest." Unpublished manuscript. JMCH.
1980h. "Sexual Democracy, American Birthright." *Best of Forum,* winter: 4-5.
1980i. Review of *Transsexuality in the Male,* by Erwin K. Koranyi. *SIECUS Report* 9, 1: 16-17.
1981a. "Comment on Transsexualism." [Note: "written for Sessuologia, but not used."] February. JMCK.
1981b. "Interview with Lynne Agress." *Pillow Talk* 2, 11: 82-88.
1981c. "Male Sexuality in 1990." In *Goals in Male Reproductive Research,* edited by Saul Boyarsky and Kenneth Polakoski, 53-59. New York: Pergamon Press.
1981d. "Sexosophy and Sexology, Philosophy and Science: Two Halves, One Whole." In *Sexology, Sexual Biology, Behavior and Therapy,* edited by Z. Hoch and H.I. Lief. Amsterdam: Excerpta Medica. *VP,* 568-79.
1981e. "A Trans-Sexual's Case History." *British Journal of Sexual Medicine* 8: 62.
1981f. Review of *Turner's Syndrome: A Psychiatric-Psychological Study of 45 Women with Turner's Syndrome,* by J. Nielsen et al. *Journal of Child Psychology and Psychiatry* 22: 100.
1982a. "Lesbian Lizards." *Psychoneuroendocrinology* 7, 2-3: 155.
1982b. "To *Quim* and to *Swive*: Linguistic and Coital Parity, Male and Female." *Journal of Sex Research* 18, 2: 173-76.
1982c. "Sex, Taboo and the Media." *Forum* 12, 1: 41-43.
1983a. [and Jackie Davison]. "Adult Penile Circumcision: Erotosexual and Cosmetic Sequelae." *Journal of Sex Research* 19, 3: 289-92.
1983b. [and G.K. Lehne]. "Biomedical and Criminal-Justice Concepts of Paraphilia: Developing Convergence." *Medicine and Law* 2: 257-61.
1983c. "Forensic Sexology." *Medicine and Law* 2: 157-58.
1983d. "The Genealogical Descent of Sexual Psychoneuroendocrinology from Sex and Health Theory: The Eighteenth to the Twentieth Centuries." *Psychoneuroendocrinology* 8, 4: 391-400.
1983e. "New Phylism Theory and Autism: Pathognomic Impairment of Troopbonding." *Medical Hypotheses* 11: 245-50. *VP,* 123-39.
1983f. [and Gregory K. Lehne and Bernard F. Norman]. "Psychology of Syndromes: IQ and Micropenis." *American Journal of Diseases of Children* 137: 1083-86.
1983g. "Sex Offending: Law, Medicine, Media and the Diffusion of Sexological Knowledge." *Medicine and Law* 2: 249-55.
1984a. "Five Universal Exigencies, Indications and Sexological Theory." *Journal of Sex and Marital Therapy* 10: 229-38. *VP,* 108-20.
1984b. "Paraphilias: Phenomenology and Classification." *American Journal of Psychotherapy* 38, 2: 164-78. *VP,* 439-53.
1985a. "The Conceptual Neutering of Gender and the Criminalization of Sex." *Archives of Sexual Behavior* 14: 279-90. *VP,* 591-600.
1985b. *The Destroying Angel: Sex, Fitness and Food in the Legacy of Degeneracy Theory, Graham Crackers, Kellogg's Corn Flakes and American Health History.* Buffalo, NY: Prometheus Books.
1986a. Original transcript of interview by Kathleen Stein for *Omni* magazine. JMCH.
1986b. "Interview" [by Kathleen Stein]. *Omni* magazine 8, 7: 78-82.

1986c. *Lovemaps: Clinical Concepts of Sexual/Erotic Health and Pathology, Paraphilia, and Gender Transposition in Childhood, Adolescence, and Maturity*. Buffalo, NY: Prometheus Books.

1986d. *Venuses Penuses: Sexology, Sexosophy and Exigency Theory*. Buffalo, NY: Prometheus Books.

1987a. "Endocrinologist: Adolescence Is Cultural 'Invention.'" [Interview]. *Adolescent Medicine* 14, 11: 1.

1987b. Speech to National Institute of Child Health and Human Development, twenty-fifth anniversary. Recording held by Kinsey Institute.

1988a. *Gay, Straight, and In-Between: The Sexology of Erotic Orientation*. New York: Oxford University Press, 1988.

1988b. [and Margaret Lamacz]. *Vandalized Lovemaps: Paraphilic Outcome of Seven Cases in Pediatric Sexology.* Buffalo, NY: Prometheus Books.

1989. "Paleodigms and Paleodigmatics: A New Theoretical Construct Applicable to Munchausen's Syndrome by Proxy, Child-Abuse Dwarfism, Paraphilia, Anorexia Nervosa, and Other Syndromes." *American Journal of Psychotherapy* 43, 1: 15-24.

1990a. "Aural Sex: Telephone Smut for Fun and Profit." [Interviewed by Max Weiss]. *City Paper* [Baltimore], July 6-12, 14, 27.

1990b. "*Cosmo* Talks to John Money Pioneer Sex Researcher." *Cosmopolitan,* December: 108-11.

1990c. "Exploring Human Sexuality." [Interviewed by Randi Henderson]. *Baltimore Sun,* March 25, 5H.

1990d. "Johns Hopkins Hospital Has a Sex Problem." [Interviewed by Frank Kuznik]. *Baltimore Magazine,* September: 42-45, 78.

1990e. "Sex: The Good, the Bad and the Kinky." *Playboy,* July: 45-49.

1990f. "Silent Shame." [Interviewed by Richard Krawiec]. *Pittsburgh Magazine,* May: 47-50.

1991a. "Explorations in Human Behavior." In *The History of Clinical Psychology in Autobiography,* vol. II, edited by C. Eugene Walker, 231-71. Pacific Grove, CA: Brooks/Cole.

1991b. "You Are Living in the Next Sexual Dark Age: Dr. John Money on the Sexual Counter-Reformation and How It Is Going to Foul-Up Our Children." *Harry* [Baltimore]: 3, 6-7.

1992a. *The Kaspar Hauser Syndrome of "Psychosocial Dwarfism": Deficient Statural, Intellectual, and Social Growth Induced by Child Abuse.* Buffalo, NY: Prometheus Books.

1992b. "Speech to National Commission on AIDS, New Orleans, May 18-19, 1992." *Kansas Biology Teacher* 3: 20.

1992c. [and V.N. Joshi and K.S. Prakasam]. "Transsexualism and Homosexuality in Sanskrit: 2.5 Millennia of Ayurvedic Sexology." *Gender Dysphoria* 1, 2: 32-34.

1993a. *The Adam Principle: Genes, Genitals, Hormones, and Gender: Selected Readings in Sexology*. Buffalo, NY: Prometheus Books.

1993b. "Parable, Principle, and Military Ban." *Society* 31, 1: 22-23.

1994. "Body-Image Syndromes in Sexology: Phenomenology and Classification." *Journal of Psychology and Human Sexuality* 6, 3: 31-48.

1995. *Gendermaps: Social Constructionism, Feminism, and Sexosophical History.* New York: Continuum.

1996. "Breakthrough Medicine." *Omni,* online chat, October 23.

1997. *Principles of Developmental Sexology*. New York: Continuum.
1998a. "Case Consultation: Ablatio Penis." *Medicine and Law* 17, 1: 113-23.
1998b. *Sin, Science, and the Sex Police: Essays on Sexology and Sexosophy*. Buffalo, NY: Prometheus Books.
1999a. *The Lovemap Guidebook*. New York: Continuum.
1999b. *Unspeakable Monsters in All Our Lives*. Amherst, NY: Prometheus Books.
2002a. *A First Person History of Pediatric Psychoendocrinology*. New York: Kluwer Academic/Plenum Publishers.
2002b. "Once Upon a Time I Met Alfred C. Kinsey." *Archives of Sexual Behavior* 31, 4: 319-22.

Works by Other Authors

Andrews, Dr. George. 1971. Letter, February 24. JMCK.
Angelides, Steven. 2004. "Feminism, Child Sexual Abuse, and the Erasure of Child Sexuality." *GLQ: A Journal of Lesbian and Gay Studies* 10, 2: 141-77.
Angier, Natalie. 2000. "X + Y = Z." *New York Times*, February 20, 10.
Atkinson, C.J. 1909. "The Boy Problem." *Canadian Club*, December 20, 52-60. http://www.canadianclub.org/Events/EventDetails.aspx?id=209 8/30/2011.
Bancroft, John. 1991. "John Money: Some Comments on His Early Work." In *John Money: A Tribute*, edited by Eli Coleman, 1-8. Binghamton, NY: Haworth Press.
Bergler, Edmund. 1956. *Homosexuality: Disease or Way of Life?* New York: Hill and Wang.
Berlin, Fred S., and Edgar Krout. 1986. "Pedophilia: Diagnostic Concepts, Treatment, and Ethical Considerations." In *Out of Harm's Way: Readings on Child Sexual Abuse, Its Prevention and Treatment*, edited by Dawn C. Haden, 155-71. Westport, CT: Oryx Press.
Bradley, Susan, Gillian D. Oliver, Avinoam B. Chernick, and Kenneth J. Zucker. 1998. "Experiment of Nurture: Ablatio Penis at 2 Months, Sex Reassignment at 7 Months, and a Psychosexual Follow-Up in Young Adulthood." *Pediatrics* 102, 1: e9.
Bull, Chris. 2000. "Body Politics." *Washington Post*, "Book World" column, April 30, X09.
Bullough, Vern L. 2003. "The Contributions of John Money: A Personal View." *Journal of Sex Research* 40, 3: 230-36.
Butler, Judith. 2001. "Doing Justice to Someone: Sex Reassignment and Allegories of Transsexuality." *GLQ* 7, 4: 621-36.
–. 2004. *Undoing Gender*. New York: Routledge.
Chomsky, Noam. 1965. *Aspects of the Theory of Syntax*. Cambridge, MA: MIT Press.
Colapinto, John. 1997. "The True Story of John/Joan." *Rolling Stone*, December 11: 54-73.
–. (2000) 2006. *As Nature Made Him: The Boy Who Was Raised as a Girl*. New York: Harper.
Coleman, Eli, ed. 1991. *John Money: A Tribute*. Binghamton, NY: Haworth Press.
Condon, Bill, dir. 2004. *Kinsey*. Fox Searchlight.
Cox, Jeffrey, and Shelton Stromquist, eds. 1998. *Contesting the Master Narrative: Essays in Social History*. Iowa City: University of Iowa Press.
Craig, Olga. 2004. "Neither Boy Nor Girl." *Calgary Herald*, March 7, B4.

Diamond, Milton. 1965. "A Critical Evaluation of the Ontogeny of Human Sexual Behavior." *Quarterly Review of Biology* 40, 2: 147-75.
–. 1998a. "Bisexuality: A Biological Perspective." In *Bisexualities: The Ideology and Practice of Sexual Contact with Both Men and Women,* edited by Erwin J. Haeberle and Rolf Gindorf, 53-80. New York: Continuum.
–.1998b. "Duke of Dysfunction." *New Zealand Listener,* September 5, 45.
–. 2000a."The John/Joan Case: Another Perspective." http://www.hawaii.edu/PCSS/biblio/articles/2000to2004/2000-another-perspective.html.
–. 2000b. "Sex and Gender: Same or Different?" *Feminism and Psychology* 10, 1: 46-54.
–. 2011. "Developmental, Sexual and Reproductive Neuroendocrinology: Historical, Clinical and Ethical Considerations." *Frontiers in Neuroendocrinology* 32, 2: 255-63.
–. 2012. "Intersex and Transsex: Atypical Gender Development and Social Construction." *Women's Studies Review* 19: 76-91.
–. 2013. "Transsexuality among Twins: Identity Concordance, Transition, Rearing, and Orientation." *International Journal of Transgenderism* 14, 1: 24-38.
Diamond, Milton, and H.K. Sigmundson. 1997. "Sex Reassignment at Birth: A Long Term Review and Clinical Implications." *Archives of Pediatric and Adolescent Medicine* 150: 298-304.
Doidge, Norman. 2000. "Hidden Agendas: Have Alfred Kinsey and John Money Set the Stage for a Backlash against Sound Sex Research?" *National Post,* February 23, B1.
Dreger, Alice Domurat. 1998. *Hermaphrodites and the Medical Invention of Sex.* Cambridge, MA: Harvard University Press.
Ehrhardt, Anke A. 2007. "John Money, Ph.D." *Journal of Sex Research* 44, 3: 223-24.
Elliott, Carl. 2000. "A New Way to Be Mad." *Atlantic Monthly,* December: 72-84.
Ellis, Albert. 1945. "The Sexual Psychology of Human Hermaphrodites." *Psychosomatic Medicine* 7: 108-25.
Eugenides, Jeffrey. 2003. *Middlesex.* Toronto: Vintage.
Farber, Susan. 1981. *Identical Twins Reared Apart: A Reanalysis.* New York: Basic Books.
Fausto-Sterling, Anne. 2000. *Sexing the Body: Gender Politics and the Construction of Sexuality.* New York: Basic Books.
–. 2012. *Sex/Gender: Biology in a Social World.* New York: Routledge.
Feinberg, Leslie. 1993. *Stone Butch Blues: A Novel.* Ithaca, NY: Firebrand Books.
–. 1996. *Transgender Warriors: Making History from Joan of Arc to RuPaul.* Boston: Beacon Press.
Fine, Cordelia. 2010. *Delusions of Gender: How Our Minds, Society, and Neurosexism Create Difference.* New York: W.W. Norton.
Fleming, Michael, Carol Steinman, and Gene Bocknek. 1980. "Methodological Problems in Assessing Sex-Reassignment Surgery: A Reply to Meyer and Reter." *Archives of Sexual Behavior* 9: 451-56.
Foucault, Michel. 1990. *The History of Sexuality: An Introduction.* Trans. Robert Hurley. New York: Vintage Books.
Frame, Janet. 1962. *The Edge of the Alphabet.* New York: G. Braziller.
–. 1984. *An Angel at My Table: An Autobiography.* London: Women's Press.
"Frank Money, 1885-1929." N.d. *Moneys Creek.* http://www.moneyscreek.co.nz/pages/frankmoney.html.
Freud, Sigmund. 2000. *Three Essays on the Theory of Sexuality.* Trans. James Strachey. New York: Basic Books.

Glassberg, K.I. 2000. "Reply." *Journal of Urology* 163: 926.
Goldie, Terry. 1989. *Fear and Temptation: The Image of the Indigene in Canadian, Australian, and New Zealand Literatures.* Montreal: McGill-Queen's University Press.
–. 2008. *Queersexlife: Autobiographical Thoughts on Sexuality, Gender and Identity.* Vancouver: Arsenal Pulp Press.
Gottman, John. 1999. *The Seven Principles for Making Marriage Work.* New York: Crown Publishers.
Green, Richard. 2008. "The Three Kings: Harry Benjamin, John Money, Robert Stoller." *Archives of Sexual Behavior* 38, 4: 610-13.
Greene, Graham. 1935. *England Made Me*. London: Heinemann, 1960.
Gulli, Cathy. 2014. "Boys Will Be Girls." *Maclean's,* January 20: 38-42.
Hacking, Ian. 1990. *The Taming of Chance.* Cambridge, UK: Cambridge University Press.
Halberstam, Judith. 2011. *The Queer Art of Failure*. Durham, NC: Duke University Press.
Hallett, Darcy, Michael J. Chandler, and Christopher E. Lalonde. 2007. "Aboriginal Language Knowledge and Youth Suicide." *Cognitive Development* 22: 392-99.
Hamer, Dean, and Peter Copeland. 1994. *The Science of Desire: The Search for the Gay Gene and the Science of Behavior.* New York: Simon and Schuster.
Hampson, John L., and Joan G. Hampson. 1961. "The Ontogenesis of Sexual Behavior in Man." In *Sex and Internal Secretion,* edited by William C. Young, 1401-32. Baltimore: Williams and Wilkins.
Hannon, Gerald. 1977-78. "Men Loving Boys Loving Men." *Body Politic* 39.
Heidenry, John. 1997. *What Wild Ecstasy: The Rise and Fall of the Sexual Revolution.* New York: Simon and Schuster.
–. 2000. "The Lynching of Dr. John Money." Was posted at http://www.theposition.com/takingpositions/ripper/00/07/31/ripper/default.shtm 7/19/00.
Herzer, Manfred. 1986. "Kertbeny and the Nameless Love." *Journal of Homosexuality* 12, 1: 1-25.
Holmes, Morgan. 2008. *Intersex: A Perilous Difference*. Selinsgrove, PA: Susquehanna University Press.
Hooker, Evelyn. 1965. "An Empirical Study of Some Relations between Sexual Patterns and Gender Identity in Male Homosexuals." In *Sex Research: New Developments,* edited by John Money, 24-52. New York: Holt, Rinehart and Winston.
Hubbard, Thomas K., and Beert Verstraete. 2013. *Censoring Sex Research: The Debate over Male Intergenerational Relations*. Walnut Creek, CA: Left Coast Press.
Imperato-McGinley J., L. Guerrero, T. Gautier, and R.E. Peterson. 1974. "Steroid 5alpha-Reductase Deficiency in Man: An Inherited Form of Male Pseudohermaphroditism." *Science* 186: 1213-55.
Intersex Society of North America. 2008. http://www.isna.org.
–. 1976. Reply to John Money. *Science* 191, 4229: 872.
Karkazis, Katrina. 2008. *Fixing Sex: Intersex, Medical Authority, and Lived Experience.* Durham, NC: Duke University Press.
Kemp, Ross. 2008. *Gangs*. London: Penguin.
Kennedy, Hubert. 1988. *Ulrichs: The Life and Works of Karl Heinrich Ulrichs, Pioneer of the Modern Gay Movement*. Boston: Alyson.
Kessler, Suzanne J. 1998. *Lessons from the Intersexed.* New Brunswick, NJ: Rutgers University Press.

King, Michael. 1998. "The Duke of Dysfunction." *New Zealand Listener*, April 4: 18-21.
–. 2000a. "Doctored Account." *New Zealand Listener*, June 3: 46.
–. 2000b. *Wrestling with the Angel: A Life of Janet Frame*. Washington: Counterpoint.
–, ed. 1988. *One of the Boys: Changing Views of Masculinity in New Zealand*. Auckland: Heinemann.
–. 2006. *Splendours of Civilisation: The John Money Collection at the Eastern Southland Gallery*. Gore, NZ: Eastern Southland Gallery.
Kinsey, Alfred C., Wardell P. Pomeroy, and Clyde E. Martin. 1948. *Sexual Behavior in the Human Male*. Philadelphia: W.B. Saunders.
–. 1953. *Sexual Behavior in the Human Female*. Philadelphia: W.B. Saunders.
Kitchen, Martin. 2001. *Kaspar Hauser: Europe's Child*. London: Palgrave Macmillan.
Lehne, Gregory K. 2009. "Phenomenology of Paraphilia: Lovemap Theory." In *Sex Offenders: Identification, Risk Assessment, Treatment, and Legal Issues*, edited by F.M. Saleh, A.J. Grundzinskas, and J.M. Bradford, 12-24. New York: Oxford University Press.
LeVay, Simon. 2011. *Gay, Straight, and the Reason Why: The Science of Sexual Orientation*. New York: Oxford University Press.
Maccoby, Eleanor. 1974. *The Psychology of Sex Differences*. Stanford: Stanford University Press.
Mason, Nicholas. 1981. "A Trans-Sexual's Case History." *British Journal of Sexual Medicine* 8: 60-61.
Masters, William H., and Virginia E. Johnson. 1966. *Human Sexual Response*. Boston: Little Brown.
Mead, Margaret. 1971. "The Role of Sexuality in the Social Integration of Children." In *Sex in Childhood*, 85-104. Tulsa, OK: Children's Medical Center.
Medical World News. 1971. "Med-School 'Stags': Students Learn about Sex from Pornographic Films." March 12.
Meehan, Brian T. 1992. "Scholars Dispute Accuracy of OCA's Anti-Gay Video." *Oregonian*, October 8, A17.
Melville, Herman. 1967. *"Billy Budd, Sailor" and Other Stories*. New York: Penguin.
Mendelsohn, Benjamin. 1963. "The Origin of the Doctrine of Victimology." *Excerpta Criminologica* 3, 3: 239-44.
Meyer, Jon K., and Donna J. Reter. 1979. "Sex Reassignment: Follow-Up." *Archives of General Psychiatry* 36, 9: 1010-15.
Meyerowitz, Joanne J. 2002. *How Sex Changed: A History of Transsexuality*. Cambridge, MA: Harvard University Press.
Mezey, Susan Gluck. 2007. *Queers in Court: Gay Rights Law and Public Policy*. Lanham, MD: Rowan and Littlefield.
Minto, Catherine L., Lih-Mei Liao, Christopher R.J. Woodhouse, Phillip G. Ransley, and Sarah M. Creighton. 2003. "The Effect of Clitoral Surgery on Sexual Outcome in Individuals Who Have Intersex Conditions with Ambiguous Genitalia: A Cross-Sectional Study." *Lancet* 361: 1252-57.
Muecke, Stephen. 2008. "A Chance to Hear a Nyinga Song." In *Joe in the Andamans and Other Fictocritical Stories*, 80-93. Sydney: Local Consumption.
Namaste, Vivian K. 2005. *Sex Change, Social Change: Reflections on Identity, Institutions and Imperialism*. Toronto: Women's Press.
Oxford English Dictionary Online. http://www.oed.com/.
Paglia, Camille. 1993. "A Question of Sex." Introductory remarks to lecture at Johns Hopkins University, September 28. [Her comment about Money being one of

her principal intellectual influences and a leading sexologist does not have a precise attribution but is likely from this lecture.]
Preves, Sharon E. 2003. *Intersex and Identity: The Contested Self.* New Brunswick, NJ: Rutgers University Press.
Prosser, Jay. 1998. *Second Skins: The Body Narratives of Transsexuality.* New York: Columbia University Press.
Raymond, Janice. (1979) 1994. *The Transsexual Empire: The Making of the She-Male.* New York: Teachers College Press.
Reddy, Gayatri. 2005. *With Respect to Sex: Negotiating Hijra Identity in South India.* Chicago: University of Chicago Press.
Reis, Elizabeth. 2009. *Bodies in Doubt: An American History of Intersex.* Baltimore, MD: Johns Hopkins University Press.
Ritchie, James E. 1967. "Ernest Beaglehole 1906-1965." *American Anthropologist* 69, 1: 68-70.
Rogers, Donald P. 1958. "The Philosophy of Taxonomy." *Mycologia* 50, 3: 326-32.
Rolling Stone. 1990. "Hot List of 1990." May 17: 112.
Roscoe, Will. 2000. *Changing Ones: Third and Fourth Genders in Native North America.* New York: Palgrave Macmillan.
Rosin, Hanna. 2008. "A Boy's Life." *Atlantic Monthly,* November. http://www.theatlantic.com.
–. 2012. *The End of Men.* New York: Riverhead Books.
Rudacille, Deborah. 2005. *The Riddle of Gender: Science, Activism and Transgender Rights.* New York: Pantheon Books.
Sagarin, Edward. 1973. Review of *Man and Woman, Boy and Girl,* by John Money. *Journal of Sex Research* 9, 3: 271-75.
Sánchez, Francisco J., Stefanie T. Greenberg, William Ming Liu, and Eric Vilain. 2009. "Reported Effects of Masculine Ideals on Gay Men." *Psychology of Men and Masculinity* 10, 1: 73-87.
Sandfort, Theo. 1987. *Boys on Their Contacts with Men: A Study of Sexually Expressed Friendships.* Elmhurst, NY: Global Academic.
Segell, Michael. 1996. "The Phallus Fallacy." *Esquire,* December: 49.
Shapin, Steven, and Simon Schaffer. 1985. *Leviathan and the Air-Pump: Hobbes, Boyle, and the Experimental Life: Including a Translation of Thomas Hobbes.* Princeton, NJ: Princeton University Press.
Smith, Deborah. 2008. "Sex Reassignment." *Sydney Morning Herald,* October 27.
Socarides, Dr. Charles. 1971. Letter to *Medical Tribune,* March 1. JMCK.
–. 1995. *Homosexuality: A Freedom Too Far. A Psychoanalyst Answers 1000 Questions about Causes and Cure and the Impact of the Gay Rights Movement on American Society.* Phoenix, AZ: Adam Margrave Books.
Solomon, R.L. 1980. "The Opponent-Process Theory of Acquired Motivation: The Costs of Pleasure and the Benefits of Pain." *American Psychologist* 35, 8: 691-712.
Solway, Larry. 1971. *The Day I Invented Sex.* Toronto: McClelland and Stewart.
Spade, Dean. 2012. *Normal Life: Administrative Violence, Critical Trans Politics and the Limits of Law.* Cambridge, MA: South End Press.
Stark, Jill. 2009. "Gender Setters: When Doctors Play God." *Sydney Morning Herald,* May 31.
Stekel, Wilhelm. 1924. *Disorders of the Instincts and the Emotions: The Parapathic Disorders.* New York: Boni and Liveright Publishing Corporation.

Stoller, Robert. 1968. *Sex and Gender: On the Development of Masculinity and Femininity*. London: Hogarth.

Swanson, David. 1960. "Suicide in Identical Twins." *American Journal of Psychiatry* 116: 934-35.

Tennov, Dorothy. 1979. *Love and Limerence*. Lanham, MD: Scarborough House.

Time. 1966. "A Body to Match the Mind." December 2: 52.

–. 1973. "Biological Imperatives." January 8: 34.

Travis, Lawrence F., III. 2008. *Introduction to Criminal Justice*. Newark, NJ: Matthew Bender.

Virtue, Noel. 1996. *Once a Brethren Boy*. Auckland, NZ: Vintage.

Walling, William H. 1902. *Sexology*. Philadelphia: Puritan Publishing.

Ward, Phyllis. 2000. *Is It a Boy or a Girl?* [Petaluma, CA]: Intersex Society of North America.

Willard, Elizabeth Osgood Goodrich. 1867. *Sexology as the Philosophy of Life: Implying Social Organization and Government*. Chicago: J.R. Walsh.

Winter, Kathleen. 2010. *Annabel*. Toronto: House of Anansi.

Wittig, Monique. 1992. *The Straight Mind and Other Essays*. Boston: Beacon Press.

Wolf, Naomi. 2012. *Vagina: A New Biography*. New York: HarperCollins.

"Yogyakarta Principles on the Application of International Human Rights Law in Relation to Sexual Orientation and Gender Identity." 2012. http://www.yogyakartaprinciples.org.

Young, Hugh Hampton. 1937. *Genital Abnormalities, Hermaphroditism and Related Adrenal Diseases*. Baltimore, MD: Williams and Wilkins.

Zucker, Kenneth J. 1999. "Intersexuality and Gender Identity Differentiation." *Annual Review of Sex Research* 10: 1-69.

Zuger, Bernard. 1970. "Gender Role Determination: A Critical Review of the Evidence from Hermaphroditism." *Psychosomatic Medicine* 32, 5: 449-63.

Index

M

S

Printed and bound in Canada by Friesens

Set in Trajan and Minion by Artegraphica Design Co. Ltd.

Copy editor: Joanne Muzak

Proofreader: Dallas Harrison

Indexer: Cheryl Lemmens